Vibration Analysis and Control - New Trends and Developments

Vibration Analysis and Control - New Trends and Developments

Editor

Jeff Winters

scitus
academics

Vibration Analysis and Control - New Trends and Developments
Edited by **Jeff Winters**

ISBN: 978-1-68117-660-4
Library of Congress Control Number: 2016946527

616, Corporate Way, Suite 2, 4766,
Valley Cottage, NY 10989

www.scitusacademics.com

Printed in United States of America on Acid Free Paper

Preface

People became interested in vibration when the first musical instruments, probably drums or whistles, were created by them. Since then, both musicians and philosophers have been searching for the rules and laws of sound production, using them in the improvement of musical instruments and passing them future generations. Vibration refers to the periodic back-and-forth movement of the molecules of an elastic body or perhaps a medium, usually resulting in virtually any body system being displaced from its condition of equilibrium and being able to respond to the forces that are likely to restore balance. There are two categories of vibration: forced and free. Free vibrations occur when the system is disturbed and is temporarily allowed to move without any restriction. The vibrations of a spring are related to a particularly simple type called a simple harmonic motion (SHM). Forced vibration occurs when an external agency continually drives the system. A basic example is a swing of children, which is pushed on each swing. Systems undergoing SHM and driven by sinusoidal forcing are of particular interest. The purpose of vibration analysis (VA) in an industrial or perhaps maintenance environment is to reduce maintenance costs and downtime of equipment by detecting equipment defects. VA is a vital component of the Condition Monitoring program and is known as predictive maintenance (PdM). VA is usually used to detect rotating equipment defects (fans, motors, pumps and gearboxes, etc.) such as imbalance, misalignment, rolling element bearing defects and resonance conditions.

Vibration Analysis and Control - New Trends and Developments

This title imparts comprehensive information on vibration analysis and control in a variety of important fields. The book covers a wide range of research topics related to the design of passive, active and semi-active vibration control systems, vibration control devices, fault detection, finite element analysis and other current studies and applications in this exciting vibration analysis and control field.

The book will be invaluable to practitioners and researchers in this field, along with graduates and other experts and postgraduate students and newcomers, who are looking for additional information on the state of the art, challenge open problems, new trends, forward-looking solutions and developments in this field.

Table of Contents

CHAPTER 1

Optimal vibration control and co-design of very flexible actuated structures

S. Maraniello, R. Palacios'

ABSTRACT

The single shooting method is applied to the optimal control and combined structural and control design (co-design) of very flexible beams. The objective is to assess feasibility, advantages and limitations of an integrated design approach when dealing with actuated structures exhibiting large oscillations and, more generally, strongly nonlinear couplings. A gradient based approach is proposed for the large design space defined by both the optimal control and co-design problems. Numerical studies are presented for the case of a very flexible actuated pendulum with large rigid-body motion. The impact of local (B-splines) and global (discrete sines) set of basis functions is investigated for increasing levels of actuation authority, showing the importance of the time–frequency resolution of the parametrisation on the convergence properties and outcome quality of the process. Locking between control and structural disciplines around specific design points is found, thus highlighting the disadvantage of a sequential design approach. Simultaneous designing of control law and structure is seen, instead, to explore efficiently larger regions of the design space.

KEYWORDS

Vibration control; Optimization; Co-design; Control vector parametrisation; Single-shooting; Geometrically-nonlinear beams

1. INTRODUCTION

While actively controlled structures generally provide better vibration characteristics than non-controlled, or passive, ones, the integration of the controller usually comes late in the design process. Their design typically relies on a sequential approach: the structure is initially sized for passive response and an active control is introduced only in a later stage, after the main features of the system have been established [1]. Looking at the structural design process from the perspective of a multidisciplinary optimisation (MDO) problem, there is substantial evidence that this approach is likely to generate only sub-optimal design points. For a true optimum, structural properties and control should be designed simultaneously, thus leading to the concept of combined design or co-design[1], [2] and [3].

The advantages of a combined control and structural optimisation have long been proved in space structures design [4] and [5], robotics [6], and noise and vibration control[7] and [8]. These studies were based on a linear representation of the closed-loop system dynamics, which facilitated the approach to the design. Asada et al. [6], for instance, directly manipulated the position of the closed-loop system eigenvalues, while Rao [5] expressed the control gains in terms of the system energy properties. In this sense, the work from Onoda and Haftka [4] best fits the modern idea of combined design, as, aside from exploring a nested approach similar to that presented by Rao [5], they also optimised the system simultaneously with respect to optimiser gains and structural design parameters.

When dealing with the dynamics of structures outside the linear regime, however, integrating the design approaches becomes a much more challenging task. Common MDO architectures, in fact, rarely allow to take advantage of the fact that the control is inherently dependent on the evolution of a system in time [1]. The larger computational cost associated to a time-dependent nonlinear analysis is, therefore, not balanced by the optimisation approach, resulting in relatively inefficient architectures. Co-design studies dealing with complex structural dynamic models have been required, therefore, to reduce the analysis cost, either using metamodels [9] or limiting the size of the systems[10] and [11]. Nonlinear systems arising from the coupling of FEM with other disciplines, e.g. for aeroservoelasticity studies, have also required reduced-order models and linearised formulations [12] and [13].

Soft robotic manipulators [14], high-altitude long-endurance (HALE) aircraft wings or large horizontal axis wind turbine (HAWT) blades are all highly flexible structures in which possibly large deformations can lead to a nonlinear dynamical behaviour. As most of the critical operating conditions of these systems — both in terms of loading and stability — are associated with unsteady phenomena, an accurate representation of the coupling between structure and control system is a necessary, but not sufficient, condition to achieve optimal designs. Structural properties and control system should, in fact, be designed simultaneously while ensuring a balanced modelling fidelity between the disciplines.

Focusing on this aspect, this work aims to provide relevant guidelines for the co-design of slender, active structures. From a structural perspective, the analysis of these systems introduces nonlinearities when dealing with large geometrical deformations and couplings between rigid-body and flexible modes dynamics [15], [16] and [17]. From a control system perspective, open-loop analysis will be considered. In real life applications, the control of actuated structures needs to be addressed in the presence of disturbances, thus requiring a feedback control and a closed-loop analysis. A deterministic open-loop analysis, in which the control has full authority on the system behaviour, remains, however, a necessary first step in the design process.

Optimal control problems can be solved through an optimise-discretise approach, in which an optimality condition is enforced on the equations describing the system dynamics [18]. However, in actual applications, the system is often too complex to apply optimality principles. A common way around this problem is to parametrise the control signal (direct methods). Single and multiple shooting methods, in particular, are directly linked to single and multidisciplinary optimisation [1] and [19]: once a parametrisation is chosen, the coefficients of the parametrisation are directly handled by the optimisation algorithm and there is no formal difference, at the optimiser level, between closed-loop system gain optimisation and open-loop optimal control solution.

While single and multiple shooting have been successfully used in many optimal control problems, an understanding on how these methods may apply to the control of active, strongly nonlinear, structures is a required step to assess the feasibility of their integration in a co-design framework. A first question to address is thus *how* to find the optimal actuation. When using medium-to-high fidelity models and dealing with dynamics, in particular, a global reconstruction of the design space via a zero-order optimisation method is, in general, not a feasible solution. Several authors, however, have showed that shooting methods can successfully be used in conjunction with gradient based optimisers to define control laws [20] and [21]. In this respect, the most common approach is to use piecewise constant or linear representations to parametrise the control signal [22].

The single shooting (or control vector parametrisation, CVP) method can be integrated in a pre-existing MDO architecture with relatively little effort. This approach, however, has to be further explored when dealing with co-design problems. In particular, as the optimisation relies on a gradient based method, it is important to assess how the smoothness of the design space is impacted when moving from optimal control to combined design. This also implies an assessment on how *large* changes in the design space can be.

In this work, a coupled flexible-rigid body dynamics model, based on a geometrically exact beam description, has been embedded in an optimisation framework. Applying the CVP technique, the actuation on the structure has been written as an optimal control problem using both a local (B-spline) and a global (discrete sine series, DSS) parametrisation (Section 2). The methodology is used to control the dynamics of a very flexible pendulum, modelled as a beam in hinged configuration and exhibiting large deformations (Section 3.1). As the pendulum flexibility increases, not only larger deformations arise, but the level of coupling between rigid and flexible modes also increases. Conceptually, therefore, this problem has many analogies to that of the control of a very flexible manipulator or to the trajectory tracking of a flexible aircraft in calm air. In particular, the impact of several factors — namely the level of nonlinearities, the problem formulation and the parametrisation used — on the optimal control results have been assessed in 3.2 and 3.3.

The active system co-design problem is then addressed (Section 4). Due to the high level of coupling between control and structure, a multidisciplinary feasible (MDF) architecture is used [3] . The implication of different modelling choices for the actuation is shown and guidelines for the co-design process of structures exhibiting strong nonlinear dynamics are finally provided.

2. METHODOLOGY

The application of the single-shooting is tested for the control and co-design of very flexible, slender, partially supported structures. In particular, it is assumed that these may exhibit large deformations, i.e. comparable in magnitude with their largest dimension. This section starts, therefore, presenting a structural model with coupled rigid-flexible body dynamics (Section 2.1). The optimal control problem is then addressed, withSection 2.2 providing a brief introduction to direct methods for nonlinear optimal control: the single shooting method, and the way this can be extended to co-design, is here presented. The parametrisations implemented in this work are discussed in more detail in Section 2.3. An overview of the optimisation framework built for both the control and co-design is finally given in Section 2.4.

2.1. Rigid-flexible body dynamic model

For modelling slender, flexible structures, a geometrically exact beam model (GEBM) with coupled rigid-flexible body dynamics is used [23] and [24]. The model is here briefly described using the notation introduced in Ref. [25]; frame of references (FoRs) and relevant vectors are shown in Fig. 1. The rigid body dynamics is expressed in terms of translational (vA) and rotational (ωA) velocity vectors of a FoR attached to the body, A, with respect to the ground FoR G. Subscripts stand for the FoR in which quantities are projected.

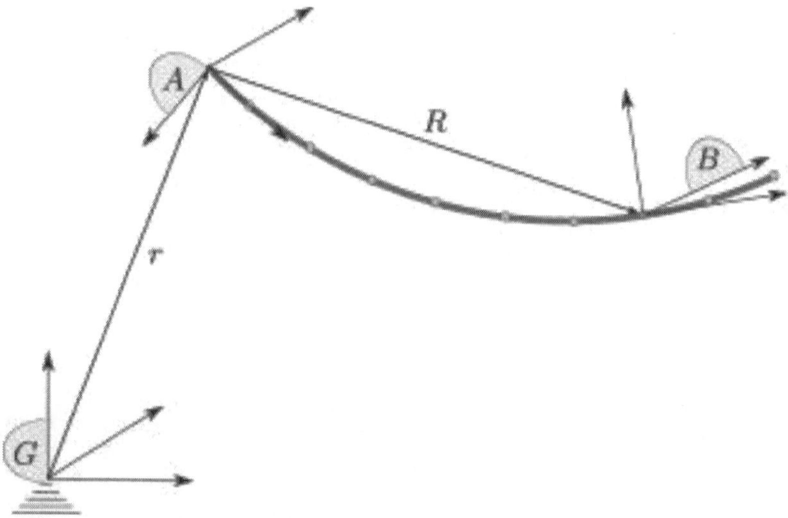

Figure. 1. Definition of frames of reference over the very flexible structure.

Local deformations are assumed to be small, thus a linear material model is used. Force and moment strains are written in terms of the position, $RA(s)$, and the Cartesian rotation vector, $\Psi(s)$, associated to a local FoR B, defined along a curvilinear coordinate s [23]. The coupled nonlinear flexible/rigid body dynamics is finally expressed using:

$$M(\eta)\begin{Bmatrix}\ddot{\eta}\\ \dot{\beta}\end{Bmatrix}+\begin{Bmatrix}Q^{s}_{\text{gyr}}(\eta,\dot{\eta},\beta)\\ Q^{r}_{\text{gyr}}(\eta,\dot{\eta},\beta)\end{Bmatrix}+\begin{Bmatrix}Q^{s}_{\text{stif}}(\eta)\\ 0\end{Bmatrix}=\begin{Bmatrix}Q^{s}_{\text{ext}}(\eta,\zeta,t)\\ Q^{r}_{\text{ext}}(\eta,\zeta,t)\end{Bmatrix}, \tag{1}$$

where $\beta^{T}=\{v_{A}^{T},\omega_{A}^{T}\}$, η is a vector containing nodal rotation and displacements, $M(\eta)$is the mass matrix, and Q_{gyr}, Q_{stif}, Q_{ext} are, respectively, gyroscopic, stiffness and external forcing terms. The latter includes control inputs and gravitational forces. For the coupled flexible-rigid body dynamics solution, the orientation of the body-attached FoR A with respect to the ground FoR G is required. This is expressed in terms of quaternions such that $\zeta^{T}=\{\zeta_{0},\zeta_{v}^{T}\}$. The scalar ($\zeta_0$) and vector ($\zeta_v$) parts of ζ are obtained via integration of body-attached FoR angular velocity, ω_A, according to [26]:

$$\dot{\zeta}_{0}=-\frac{1}{2}\omega_{A}^{T}\zeta_{v},\quad \dot{\zeta}_{v}=-\frac{1}{2}(\zeta_{0}\omega_{A}-\tilde{\omega}_{A}\zeta_{v}), \tag{2}$$

where $(\tilde{\ })$ is the skew symmetric matrix operator. Spherical joint boundary conditions (BCs) have been implemented by setting the translational velocity of the FoR A, v_A, to be zero. Hinge BCs can be derived similarly, allowing rotations only along one axis.

2.2. From optimal control to co-design

In general, an optimal control problem can be seen as an optimisation problem with a design variable, the control input u, that is a time-dependent function:

$$\begin{aligned} \text{min.} \quad & I = I(u, y, \dot{y}) \\ \text{w.r.t.} \quad & u(t), y(t) \\ \text{s.t.} \quad & c(u, y) \geq 0 \\ & R(t, u, y, \dot{y}) = 0 \end{aligned} \tag{3}$$

In problem (3)*y* is the state of the system, *I* is the cost functional to minimise and *c* are the design constraints. The set of equations *R refers* to the coupled rigid-flexible body dynamics Eqs. (1) and (2) over the time horizon [0,T]. The design constraints, $c=\{c_o,c_c\}$, are divided into a set of control specific (c_c) and general (c_o) design requirement. The first set includes constraints associated to the control input *u*: in this work, bound constraints of the form

$$\begin{aligned} u(t) \geq u_L(t) \\ u(t) \leq u_H(t) \end{aligned}, \quad t \in [0,T], \tag{4}$$

as well as initial and terminal conditions, are enforced. The second set is instead associated to any other requirement on system performance.

While analytical tools for the solution of problem (3), such as Pontryagin's maximum principle, are available, they are too complex to be used for the control of large nonlinear systems [22]. The alternative are direct methods, in which the control function is discretised in time and expressed in terms of a coefficient vector x_c, i.e. $u(t)\rightarrow u(x_c)$. In direct transcription (DT) or direct simultaneous methods, also the state is discretised in time and treated as a design variable. The optimiser, therefore, handles problem (3)directly, simultaneously solving physics and actuation. In a MDO context, this approach would be referred to as an all-at-once or simultaneous analysis and design (SAND) [3]. DT methods can explore infeasible and unstable states, thus possibly leading to faster convergence, but they introduce a large number of design variables and, in MDO problems, may also have convergence issues. If Eq. (1) is solved at each iteration for the state, problem (3) can be recast in the form of a multidisciplinary feasible architecture:

$$\begin{aligned} \text{min.} \quad & I = I(x_c, y(x_c), \dot{y}(y, x_c)) \\ \text{w.r.t.} \quad & x_c \\ \text{s.t.} \quad & c_o(x_c, y(x_c)) \geq 0 \\ & c_c(x_c, y(x_c)) \geq 0 \end{aligned} \tag{5}$$

where the dependency of the state on the control, $y=y(x_c)$, has been explicitly stated. This approach is referred to as control vector parametrisation (CVP) or single-shooting method: while a solution to Eq. (1) has to be found at each optimisation, the size of the problem is reduced to its minimum.

The extension of problem (5) to a MDO problem that simultaneously optimises control action and structural properties is straightforward and only requires the inclusion of a set of structural design parameters, x_s, and constraints, cs:

$$\begin{aligned} \text{min.} \quad & I = I(x_c, x_s, y(x_c, x_s), \dot{y}(y, x_c, x_s)) \\ \text{w.r.t.} \quad & x_c, x_s \\ \text{s.t.} \quad & c_o(x_c, x_s, y(x_c, x_s)) \geq 0 \\ & c_c(x_c, x_s, y(x_c, x_s)) \geq 0 \\ & c_s(x_c, x_s, y(x_c, x_s)) \geq 0 \end{aligned} \tag{6}$$

This is the description that will be used in this work.

2.3. Control parametrisation

In the single shooting approach, the control signal is expressed as a linear combination ofN_c basis functions, $\phi_n(t)$, defined over the time horizon [0,T]:

$$u = \sum_{n=1}^{N_c} x_{cn} \, \phi_n(t) \tag{7}$$

In most optimal control problems, a piecewise constant parametrisation is used: in addition to being easy to implement, that scheme offers good convergence properties[22]. However, to describe the movement of typical control actuators on a relatively large time domain, piecewise constant or linear parametrisations would lead to set of basis functions of substantial size. While from a computational point of view adjoint methods can deal effectively with an increased number of design variables, these present the additional development cost of building the adjoint model itself. Furthermore, the larger the basis size, the higher are the frequencies that the parametrisation can reproduce: a smoothing of the control signal to avoid unrealistically steep changes would, therefore, be required, either enforcing additional constraints on the control rate of change or smoothing the gradient itself.

With the purpose of modelling smooth actuation signals while limiting the number of coefficients used to parametrise the control, N_c, only C^1 continuous or higher parametrisations are considered for this work. As the dynamics of structures is strongly linked to the frequency of excitations of external disturbances and control forces (see also 3 and 4), it is natural to use a parametrisation that can be easily linked to the frequency range of the control. An obvious candidate is the discrete Fourier series or, for control signals with u(0)=u(T)=0, discrete sine series (DSS) obtained as

$$\phi_n(t) = \sin 2\pi f_n t \quad \text{with: } f_n = n f_0,\ f_0 = \frac{1}{2T} \tag{8}$$

The discrete Fourier series and the DSS have both the advantage of allowing a direct control of the maximum actuation frequency of the control. While sine and cosines are globally defined in time, however, they are collocated in the frequency domain, each harmonic being associated to a specific frequency f_n (see Eq. (8)). While this does not pose an issue in terms of reconstruction properties,[1] a poor frequency resolution can be a penalty during the sensitivity analysis required to solve the optimal control problem (5) via a gradient based approach. When using a DSS parametrisation, in fact, the *n* th term of the cost function gradient, dI/dx*cn*, with x*cn* defined as in Eq.(7), represents the sensitivity of the cost function *I* with respect to a change in magnitude of the *n*th harmonic parametrising the control *u*: while affecting the whole time domain, this is localised at the frequency *fn*.

The duality frequency vs. time resolution is a well known problem in signal and image processing [27]. In this sense, local basis functions can provide more flexibility in terms of capturing relevant frequency content of the structural dynamics response, particularly when large geometrical deformations imply changes in the system features (e.g. its natural frequencies) with time. To investigate this, B-splines have additionally been chosen for their smoothness properties. A set of B-splines basis functions of order *p* can be built recursively over a set of $N\tau$ control points τn as [20]:

$$\phi_n^{(0)}(t) = \begin{cases} 1 & \text{if } \tau_n < t < \tau_{n+1} \\ 0 & \text{else} \end{cases} \tag{9}$$

and

$$\phi_n^{(p)}(t) = \frac{t-\tau_n}{\tau_{n+p}-\tau_n}\phi_n^{(p-1)}(t) + \frac{\tau_{n+p+1}-t}{\tau_{n+p+1}-\tau_{n+1}}\phi_{n+1}^{(p-1)}(t), \quad p>0 \tag{10}$$

Note that, if $N\tau$ control points are used, the number of spline basis required is $N_c=N_\tau+p-1$. Third-order B-splines were found to provide good and smooth reconstructions for the applications in this work.

The frequency range of actuation can be regulated by noticing that, in order to capture a maximum frequency f_{max}, a spacing between control points $\Delta\tau=1/(2f_{max})$ is necessary (Nyquist criterion). It results that, fixed a f_{max} value, DSS and B-splines basis have comparable size and similar reconstruction properties, although it should be noted that the frequency spectrum of each spline basis has a smooth decay to zero towardsf_{max}. As it will be shown in Section 3.3, this means that harmonic components whose frequencies near f_{max} are not reconstructed accurately. While local in time, however, spline basis have a distributed frequency content, thus providing opposite properties to the DSS basis during a sensitivity analysis.

2.4. Optimisation framework

In line with Ref. [20] and [21] a standard quasi-Newton method, the SLSQP optimisation algorithm [28], has been used to solve both the nonlinear optimal control problem and the co-design optimisation. The GEBM with coupled flexible-rigid body dynamics from an in-house aeroelastic simulation environment [29] and [25] has been embedded in an optimisation framework built using OpenMDAO [30]. The implementation is monolithic and uses finite differences for the gradient evaluation.

Optimal control and co-design are solved using both DSS and B-splines. For both basis, the bound constraints in Eq. (4) are enforced by oversampling the control signal. In all the problems considered, cost and constraint functions, as well as design parameters, are scaled to achieve comparable orders of magnitude. If all the constraints are verified, the optimisation process is stopped whenever the relative change in cost and constraints is below a tolerance value of 0.1 percent.

3. NUMERICAL STUDIES ON OPTIMAL CONTROL

The methodology presented in Section 2 is verified for the control of a very flexible actuated pendulum first proposed by Wang and Yu [31]. The pendulum is modelled as a very flexible beam and can undergo large rigid body rotations and geometrical deformations. Being fully deterministic — as no external disturbances are accounted for —, this problem allows to easily assess the outcome of both the control and integrated design processes.

Both CVPs introduced in Section 2.3 are exercised for the optimal control of the system and results are compared against Ref. [31]. In order to show the effect of using different discretisations and problem formulations, the behaviour of two pendula of different flexibility is analysed (Section 3.2). The impact of the actuation bandwidth and the structural nonlinearities is further assessed in Section 3.3. Finally, a multi-resolution strategy is proposed to improve the convergence characteristics of the process (Section 3.4).

3.1. Problem description

The flexible pendulum configuration proposed in Ref. [31] is sketched in Fig. 2. The pendulum is modelled as a hinged elastic beam and lies initially in a stable equilibrium position along the vertical direction (Z-axis), gravity effects being accounted for. In order to control the system, an actuating torque, $MY(t)$, chosen to be zero at the initial and final time of the simulation, is applied at its root (Fig. 2), causing the pendulum to oscillate about the hinge point. The torque time history $MY(t)$, in particular, is optimised such as to maximise the leftward X velocity of the pendulum tip, vX, measured in the global FoR at time $T = 2$ s. The problem is fully deterministic and any sources of friction or damping are not considered.

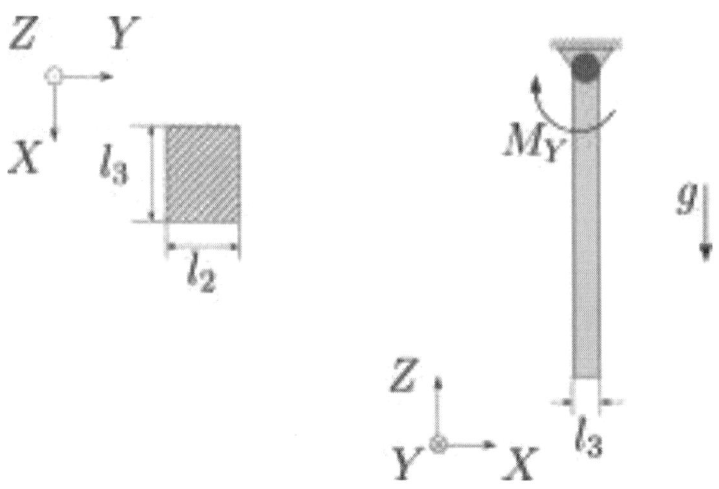

Figure. 2. Flexible pendulum geometry.

To account for the cost of actuation, Wang and Yu [31] augmented the cost function with the quadratic force regulator term

$$P[M_Y]=\frac{1}{2}\int_0^T\left[\pi_1\, M_Y^2+\pi_2\left(\frac{dM_Y}{dt}\right)^2\right]dt \tag{11}$$

which depends on both the magnitude and the rate of change of the actuation. The optimal control problem can thus be written, in its continuous form, as:

$$\begin{aligned} \text{min.}\quad & I=\kappa_1\, v_X(T)+\kappa_2\, P \\ \text{w.r.t.}\quad & M_Y(t) \\ \text{s.t.}\quad & M_Y(0)=M_Y(T)=0 \\ & -M_{max}(t)<M_Y(t)<M_{max}(t) \end{aligned} \tag{12}$$

where I is the augmented cost function and the constants κi and πi are scaling parameters to ensure that all the terms in the cost function and the force regulator equations have same units and comparable magnitude; note, in particular, that κ_1 is required to have negative value. In the following, we will refer to problem (12) as the augmented problem. In the numerical implementation, MY is discretised by means of DS (Eq. (8)) and B-spline (Eqs. (9) and (10)) basis functions; the optimisation is thus performed with respect to the coefficients of the parametrisation, xc.

The augmented problem (12) defines a trade-off between performance (vX term) and cost of actuation (P term). While this approach is in line with many practical control strategies (e.g. LQR feedback controllers), it can also affect the convexity of the design space and lead to path dependent solutions in the optimisation process. Therefore, the optimal control problem has also been formulated in a constrained version:

$$\begin{aligned} \text{min.}\quad & v_X(T) \\ \text{w.r.t.}\quad & M_Y(t) \\ \text{s.t.}\quad & P\le P_{max} \\ & M_Y(0)=M_Y(T)=0 \\ & -M_{max}(t)<M_Y(t)<M_{max}(t) \end{aligned} \tag{13}$$

The integral term P acts as a measure for the energy that the actuation can transfer into the dynamical system. Constraining this quantity allows, therefore, to bound the level of kinematic displacements displayed by the structure and, consequently, the amount of nonlinear effects experienced of the system. As it will be shown in Section 3.3, this has a significant impact on the design space smoothness.

3.2. Optimal control results

The optimal control problem for the flexible pendulum (Fig. 2) is initially solved using the augmented formulation defined in problem (12) and results are compared to those obtained by Wang and Yu [31]. The pendulum is modelled as a beam of constant rectangular cross-section with area $A = 10^{-2}\,\mathrm{m}^2$ and negligible rotational inertia; an isotropic material of Young's module $E = 1.2\,\mathrm{Pa}$ and density $\rho = 100\,\mathrm{kg\,m^{-3}}$ was used to reproduce the inertia and stiffness properties in Ref. [31]. The actuation was bounded not to exceed absolute value of $M_{max} = 3.5\,\mathrm{N\,m}$, while cost and penalty term parameters appearing in problem (12) were set as per Ref. [31] to be

$$\kappa_1 = -1\,\mathrm{s\,m^{-1}}, \quad \kappa_2 = 1$$
$$\pi_1 = 1\,\mathrm{N^{-2}\,m^{-2}\,s^{-1}}, \quad \pi_2 = 10^{-2}\,\mathrm{N^{-2}\,m^{-2}\,s}$$

Since rigid body rotations and deformations are all planar, the only relevant elastic quantity is the bending stiffness in the plane of motion. Two beams, one being 10 times stiffer than the other, are obtained varying the sides of the cross-section l_2 and l_3 while keeping the sectional area. They are referred to as *stiff* and *flexible* pendulum in Table 1.

The characteristic frequencies in Table 1 have been computed around the underformed pendulum configuration to give an idea of where the resonance points are located in the nonlinear structure. Note that, as the structure becomes more flexible, the natural frequency related to the first bending mode f_b drops and the coupling between flexible and rigid body dynamics increases.

Table 1. Pendula structural properties for the optimal control problem.

Case	l_2 (m)	l_3 (m)	EI (N m^2)	f_r (Hz)	f_b (Hz)
stiff pendulum	0.1000	0.1000	10.0	0.50	7.76
flexible pendulum	0.3162	0.0316	1.0	0.50	2.45

For both cases proposed in Table 1 the optimal control problem is solved with B-spline and DSS parametrisations of different basis sizes, based on the maximum frequency, f_{max}, captured by the control. However, fixed a certain value of f_{max}, the size of DSS and B-spline basis are comparable (Section 2.3). In particular, for each parametrisation the basis size was chosen such as to exclude ($f_{max} < f_b$, low bandwidth control) and include ($f_{max} > f_b$, high bandwidth control) the flexible mode natural frequency of vibration. It should be finally noted that the range of frequencies of interest falls within the domain of simple optical recognition methods [32], making this set-up suitable for experimental verification.

3.2.1. Stiff pendulum

The *stiff* pendulum will be subject to the control actions in Table 2. In the table, *Nc* and f_{max} refer to the basis size of each parametrisation and the related maximum frequency of actuation, respectively. Optimal control results are presented in terms of cost (*I*), penalty factor (*P*) and final pendulum tip velocity, *vX*(*T*). *NI* is the number of iterations required to complete the optimisations.

When using $f_{max} = 2$ Hz, the optimal actuation does not excite the first bending mode of the pendulum, which, therefore, mostly swings rigidly. This can be observed in Fig. 3, where the snapshots of the actuated pendulum position, driven by the DSS modelled optimal torque, show no relevant elastic deformation. Comparing optimal actuation and tip displacements time histories against Ref. [31] demonstrates that both the spline and DSS parametrisation can capture well the rigid-body motion frequency (Fig. 4), thus returning very similar performances (Table 2). As physically expected, the control moment, *MY*, excites the rigid-body motion only and uses the gravity potential energy to increase the final tip velocity, *vX*(*T*), while limiting the overall actuation cost, *P*.

Table 2. Optimal control results of the *stiff* pendulum using different parametrisations.

Parametrisation	N_c	f_{max} (Hz)	N_I	$v_X(T)$ (m s^{-1})	P	I
spline	11	2	20	6.08	2.73	−3.35
DS	8	2	14	6.08	2.73	−3.35
spline	43	10	33	12.35	6.18	−6.17
DS	40	10	44	10.57	4.77	−5.80

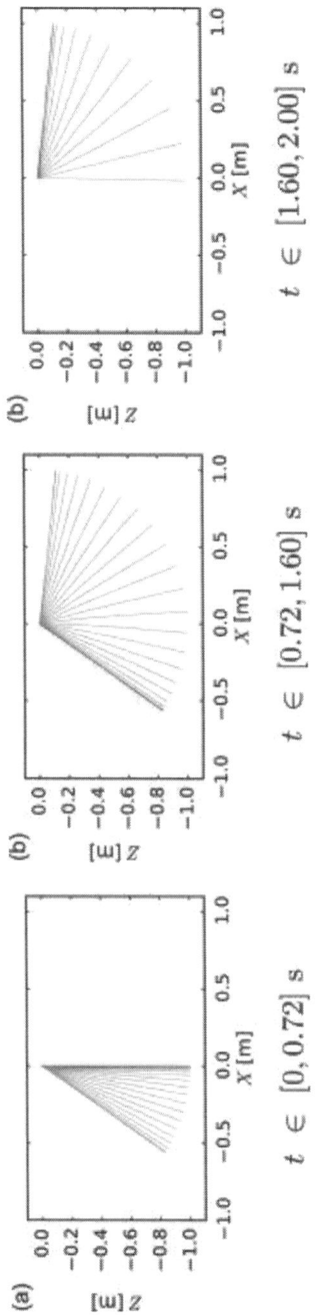

Figure. 3. Snapshots (25 frames per second) of the *stiff* pendulum response for the optimal actuation with a control maximum frequency $f_{max} = 2$ Hz using a DSS parametrisation. (A thick line identifies the initial shape).

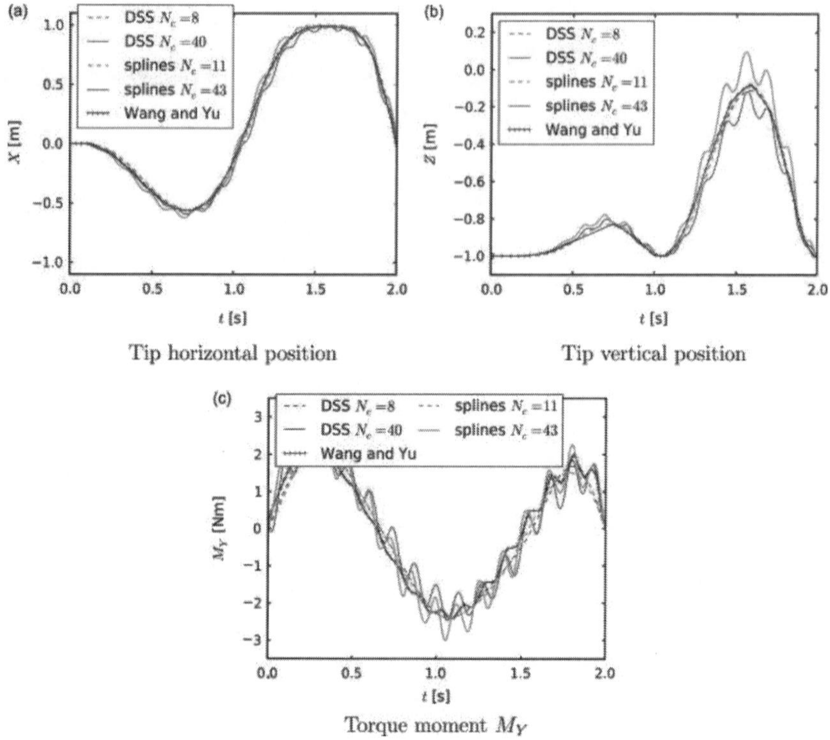

Figure. 4. Time histories under optimal control of the *stiff* pendulum using different parametrisations of the torque signal.

Setting $f_{max}>f_b$ leads to a large increase of the final tip velocity, $vX(T)$, as the first bending mode is excited (Fig. 5). The active system is now capable of storing elastic energy, which is converted into kinetic energy as t→T, providing a further contribution tovX and enhancing the overall system performance. With both splines and DSS, the high frequency component is larger than in the solution obtained by Wang and Yu [31]. It has to be noticed, however, that results in Ref. [31] are obtained from optimisations that are stopped after only 10 iterations and a beam model that includes structural damping. This likely explains the small differences between both sets of results.

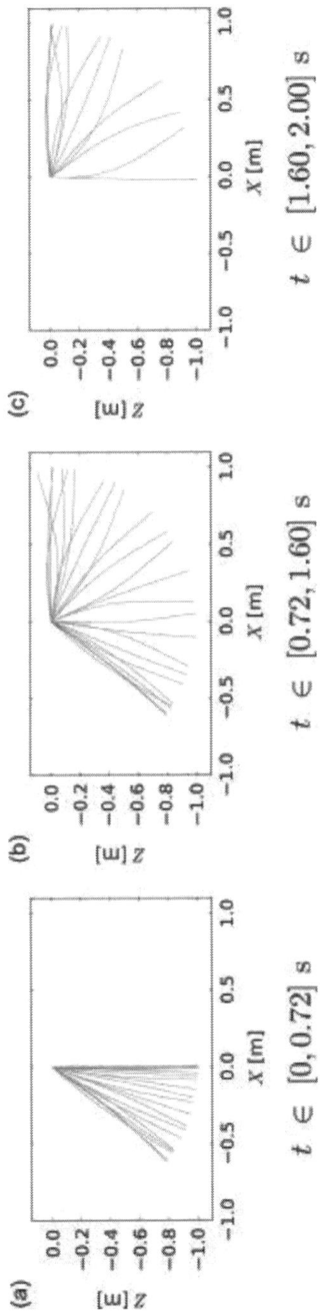

Figure. 5. Snapshots (25 fps) of the *stiff* pendulum response for the optimal actuation obtained using a B-spline parametrisation with Nc=43 control points ($f_{max} = 10$ Hz). A thick line identifies the initial shape.

While the optimal actuation always shows to correctly exploit the system physics, the 10 Hz bandwidth solutions obtained with the two parametrisations point towards two different minima, corresponding to two different levels of control actuation, *P*. To verify the existence of local minima, the control design space has been sampled for a case in which the pendulum is actuated by a control parametrised with only two sine waves. To model a low bandwidth actuation, these were chosen to have frequencies 0.5 Hz and 0.75 Hz, with amplitudes $x_{0.5}$ and $x_{0.75}$, respectively. These are associated to a left/right and left/right/left rigid-body mode and, as physically expected, had been observed to have a large contribution in both the 2 Hz and 10 Hz DSS parametrised optimal actuations.

The contour lines of the cost *I*, obtained when varying the amplitude of the two sine waves, are shown in Fig. 6a, together with the isolines of the force regulator term, *P* (thick curves). The design space is smooth with a minimum at $(x_{0.5},x_{0.75})$=(0.88,1.83) N m. Despite the reduced size of the parametrisation, cost and penalty at the minimum (I=−2.94 and *P* =2.46) compare well with those obtained using a low bandwidth actuation (f_{max} =2 Hz in Table 2).

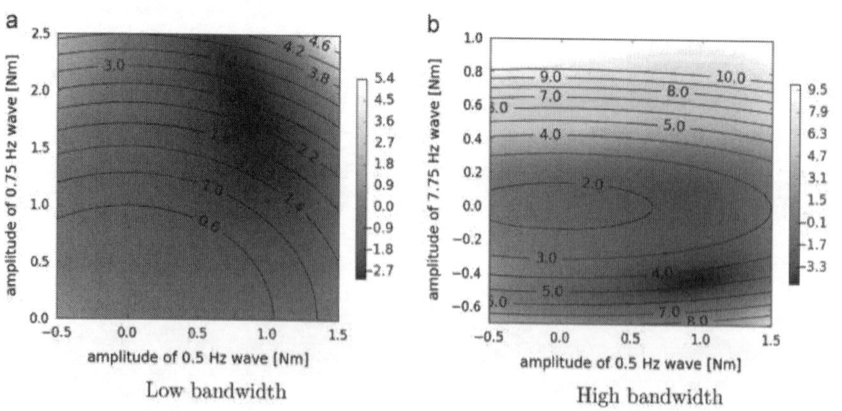

Figure. 6. Cost function, *I*, associated to a *rigid* pendulum when varying the amplitude of the sine waves parametrising the actuating torque, *MY*.

To model a high bandwidth control, the amplitude $x_{0.75}$ of the 0.75 Hz sine wave was fixed to be 1.7 N m and a high frequency sine wave (7.75 Hz, near the first natural frequency) of amplitude $x_{7.75}$ was added to the parametrisation. As before, both coefficients of the parametrisation were varied so as to reconstruct the cost function *I*over the control design space (Fig. 6b). While a minimum — corresponding to a predominantly low-frequency actuation that only exploits the rigid-body dynamics — is still observed for $x_{7.75} \approx 0$, a new minimum, corresponding to a peak of structural resonance, appears. A further investigation on how both minima change with the amplitude and frequency of the excitation is included in Section 4.1.

Sampling the cost function *I* has therefore shown that when the control can exploit resonances, multiple minima can populate the design space. The reason why the high bandwidth actuations provided by splines and DSS lead to distinct point of minimum can then be linked to how different sets of basis functions reconstruct the actuation. While a control signal parametrised with a low number of spline basis cannot be built for direct comparison with Fig. 6b, it is clear that as the mapping between parametrisation coefficients *xc* and final cost *I* changes, so does the location of the minima.

3.2.2. Flexible pendulum

The study of the low-bandwidth actuated *stiff* pendulum has shown that, when the actuation can only excite the rigid-body mode of vibration, the design space is smooth and the actuations provided by splines and DSS parametrisations are in good agreement. As the pendulum flexibility is increased, however, the distance between rigid and flexible body characteristic frequencies, *fr* and *fb*, is drastically reduced (Table 1) and it becomes harder to excite one of the modes without exciting the other.

The implications of an increasing rigid-flexible body dynamics coupling are highlighted by the results obtained using DSS and spline basis to parametrise two low-bandwidth actuations ($f_{max} < f_b$) having

$f_{max} = 1.25$ Hz and a $f_{max} = 1.5$ Hz. The optimal torque time histories (Fig. 8a) and the associated cost, *I*, and penalty factor, *P* , compare well across parametrisations only when $f_{max} = 1.25$ Hz (Table 3). However, as clearly shown by the snapshots in Fig. 7, the DSS modelled actuation with $f_{max} = 1.5$ Hz can excite the pendulum bending mode, allowing to reach a cost 22 percent lower than when using a splines parametrisation. This is verified in Fig. 8b, where the DS decomposition of the optimal actuations are shown.[2] While for the $f_{max} = 1.25$ Hz cases these can only excite the rigid-body dynamics, raising f_{max} allows the control to mildly excite the pendulum bending mode. The different resolution between the parametrisation becomes, at this point, critical: the DSS basis can express a strong frequency content around $f_{max} = 1.5$ Hz, whereas this cannot be achieved using splines. The more aggressive control obtained using DSS (note that the force regulator term *P* almost doubles with respect to the corresponding spline case) is, therefore, justified by the different resolution.

As seen in Fig. 9, when using a higher bandwidth control ($f_{max}>f_b$), both discretisations capture well the very flexible beam dynamics. However, and as already discussed for the*rigid* pendulum in Section 3.2.1; the dynamics under resonance increases the path dependency of the results. This can be seen in Fig. 8c and d: while the final cost, *I*, compares well across parametrisations, the splines obtained solution has now a considerably more aggressive actuation (Table 3). Results also include those of Ref.[31], which account for damping in the model and, as before, show a smoother behaviour.

3.2.3. Path dependency vs. problem formulation

When dealing with resonance conditions the optimal actuation torque, *MY*, obtained from the augmented problem (12), was found to be parametrisation dependent. The control of the *flexible* pendulum has also underlined that the parametrisation resolution can also have an effect, especially when dealing with an increasing coupling between flexible and rigid body dynamics. To isolate the impact of the problem formulation from that of the parametrisation (which will be the focus of Section 3.3), the optimal control of the *stiff* and*flexible* pendulum is obtained now using the constrained formulation (13). Large deformations are still allowed due to a large maximum actuation (P_{max}=6) and a high bandwidth ($f_{max}>f_b$). The results are shown in Table 4 and present a much higher consistency in the optimal control across both discretisations. Having removed the penalty term from the cost function definition, the optimal actuation always reaches, as expected, the limit P_{max}=6. For the *stiff* pendulum case, the two parametrisations are in extremely good agreement. As the system becomes more nonlinear (*flexible* pendulum), however, spline basis seem to be able to better capture the system resonance, leading to a 1.9 percent higher tip velocity $vX(T)$.

Fig. 10 shows however that better performance of the spline is not linked to better reconstruction properties. Converting the spline parametrised optimal actuation in a DSS shows, in fact, that this has no relevant frequency content over $f_{max} = 4$ Hz (Fig. 10a). In fact, the spline optimal actuation is reconstructed using a DSS series in Fig. 10b. Applying the reconstructed torque to the pendulum, a final tip velocity $v_X(T) = 15.62$ m s^{-1}, which is in line with the performance provided by the spline optimal actuation, is achieved.

It can thus be concluded that, despite the use of a *constrained* problem formulation, the two parametrisations can still converge to different minima as the nonlinearity of the system increases. While results are in very good agreement for the *stiff* pendulum, the DSS actuation leads to a control with a higher low frequencies for the *flexible* pendulum case (Fig. 10a). The use of a *constrained* problem formulation, however, reduces the gap between the two solutions, as the level of actuation, P_{max} — and consequently the final tip velocity, $vX(T)$ — are now comparable. This formulation, which diverges from that of [31], has been therefore chosen to further investigate the impact of both frequency resolution and level of actuation in the next section.

Table 3. Optimal control results of the *flexible* pendulum using different parametrisations.

Parametrisation	N_c	f_{max} (Hz)	N_l	$v_X(T)$ (m s^{-1})	P	I
spline	8	1.25	29	4.63	1.59	−3.04
DS	5	1.25	16	4.48	1.59	−2.89
spline	9	1.5	22	5.40	2.19	−3.21
DS	6	1.5	40	8.42	4.50	−3.92
spline	19	4	57	15.92	5.73	−10.19
DS	16	4	53	13.11	3.48	−9.63

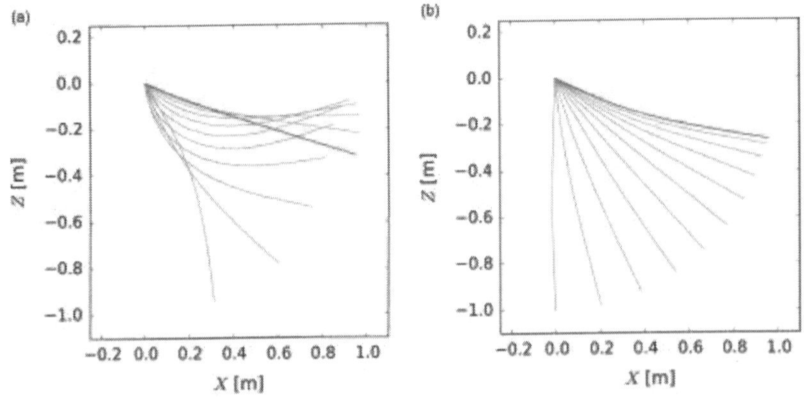

DSS, $N_c = 6$ B-splines, $N_c = 9$

Figure. 7. Snapshots (25 fps) of the *flexible* pendulum response for t∈[1.60,2.00] s with optimal control parametrisation having a maximum frequency $f_{max} = 1.5$ Hz. (A thick line identifies the initial shape).

3.3. Parametrisation vs. system nonlinearity

In order to investigate further the duality between local and global parametrisations, the optimal control problem of the pendulum is solved using both B-splines and DSS for a range of beams of different stiffness. In both cases the number of basis functions is chosen such as to achieve a maximum frequency $f_{max} = 6$ Hz. The pendulum bending stiffness is varied by changing the aspect ratio of the beam cross section (Fig. 2) under constant sectional area, $A = 10^{-2}$ m. As a result, the rigid-body mode natural frequency, f_r, remains unaltered.

Varying l_3 between [0.05,0.10] m resulted in a design space including situations for which the chosen maximum control frequency is either greater or smaller than the pendulum first bending natural frequency, f_b. More specifically, f_b has values between 3.88 Hz and 7.76 Hz and it is $f_b = f_{max}$ when EI=5.98 N m^2 and $l_3 = 0.0773$ m. The lower bound for the beam stiffness is such that the resonance of the second bending mode is avoided. While this could also be exploited, it was not included in this exploratory study to simplify the analysis.

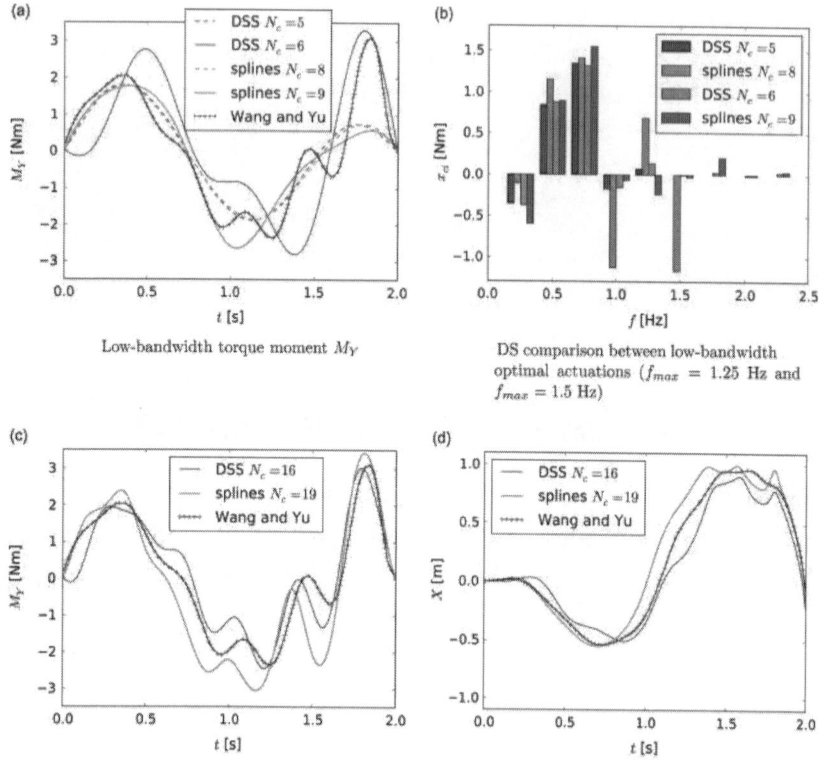

Low-bandwidth torque moment M_Y

DS comparison between low-bandwidth optimal actuations (f_{max} = 1.25 Hz and f_{max} = 1.5 Hz)

High-bandwidth torque moment M_Y

Tip horizontal position under high-bandwidth actuation

Figure. 8. Time histories under optimal control of the *flexible* pendulum using different parametrisations of the torque signal.

The constrained optimisation problem (13) is solved for three levels of admissible actuation (P_{max}=2,4,6) to assess the impact of increasing nonlinear effects on the solution. Results are shown in Fig. 11. The design space can be divided into three regions according to whether the control can or cannot excite the pendulum bending mode. When the bending natural frequency, f_b, is well below the maximum frequency of actuation f_{max} $(l_3 \lesssim 0.070\ \mathrm{m})$, the control signal can excite the rigid-flexible body dynamics, providing high performance. In this situation, in fact, the actuation has a high bandwidth in relation to the physical properties of the system ($f_{max} \gg f_b$). When $l_3 \gtrsim 0.085\ \mathrm{m}$, on the other hand, the same control with $f_{max} = 6\ \mathrm{Hz}$ has now a low bandwidth. As a result, only the rigid body dynamics is exploited and performance decreases. At last, a transition region $(0.070\ \mathrm{m} \lessapprox l_3 \lessapprox 0.085\ \mathrm{m})$, in which resonance outsets, can be identified.

The classification introduced can be more clearly understood by looking at Fig. 12, where the magnitude ($|xcn|$ in Eq. (7)) of the basis functions of the DSS parametrised optimal control is shown for a range of different pendula stiffness and actuation cost, P_{max}. As it can be seen, the control effort is concentrated in two regions, the first one located in the low frequency range so as to excite the rigid body dynamics. The second one is observed when fb is in the control range: a peak around this frequency appears and, as the beam stiffness varies, this moves to follow the bending mode natural frequency. Increasing the level of actuation, P_{max}, larger geometrical deformations appear and resonance intensifies. The consequently higher level on nonlinearities in the problem is clearly reflected in the design space smoothness, which is visibly reduced when passing fromP_{max}=2 (Fig. 12a) to P_{max}=6 (Fig. 12c).

Under low frequency actuation ($f_{max} \ll f_b$), spline basis can achieve slightly better performance because their spectrum extends above $f_{max} = 6\ \mathrm{Hz}$, allowing the torque to mildly excite the pendulum bending mode. The phenomenon is consistently stronger for larger values of actuations.

As expected, the transition region $(0.070\ \mathrm{m} \lessapprox l_3 \lessapprox 0.085\ \mathrm{m})$ sees a change in performance due to the switch from rigid-flexible to rigid body only dynamics. The difference increases with P_{max}, as stronger actuations induce a more intense level of resonance. The change in maximum tip velocity, $vX(T)$, is sharper when using a DSS, whose related optimal control generally provide higher performance than the spline one. Contrarily to what it was seen in Section 3.2.3, however, this is not linked to the DSS providing better convergence properties but to the fact that in the transition region the solution quality is dictated by the parametrisation frequency content.

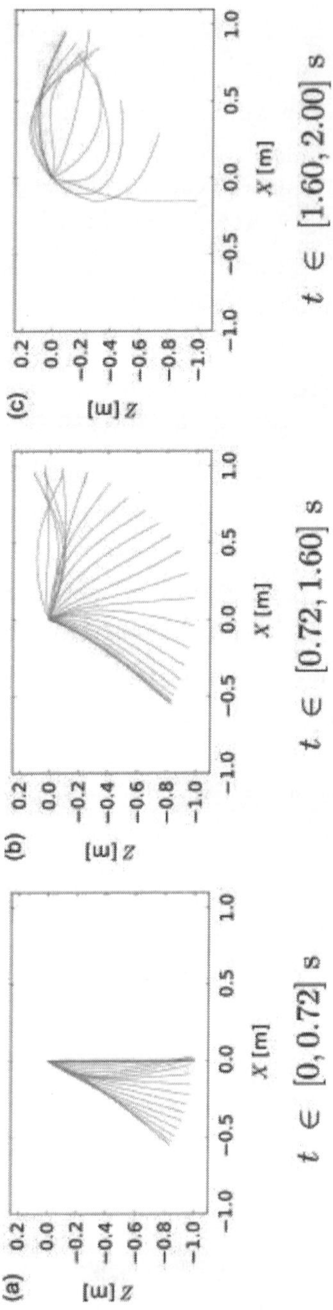

Figure. 9. Snapshots (25 fps) of the *flexible* pendulum response for the optimal actuation obtained using a B-spline parametrisation with Nc=19 control points ($f_{max} = 4$ Hz). A thick line identifies the initial shape.

Table 4. Optimal control results for the *stiff* and *flexible* pendulum using the constrained formulation (13) and settingP_{max}=6.

Pendulum	Basis	f_{max} (Hz)	N_I	$v_X(T)$ (m s^{-1})	P
stiff	spline	10	40	12.30	6.00
stiff	DS	10	170	12.27	6.06
flexible	spline	4	54	15.73	6.00
flexible	DS	4	158	15.43	5.96

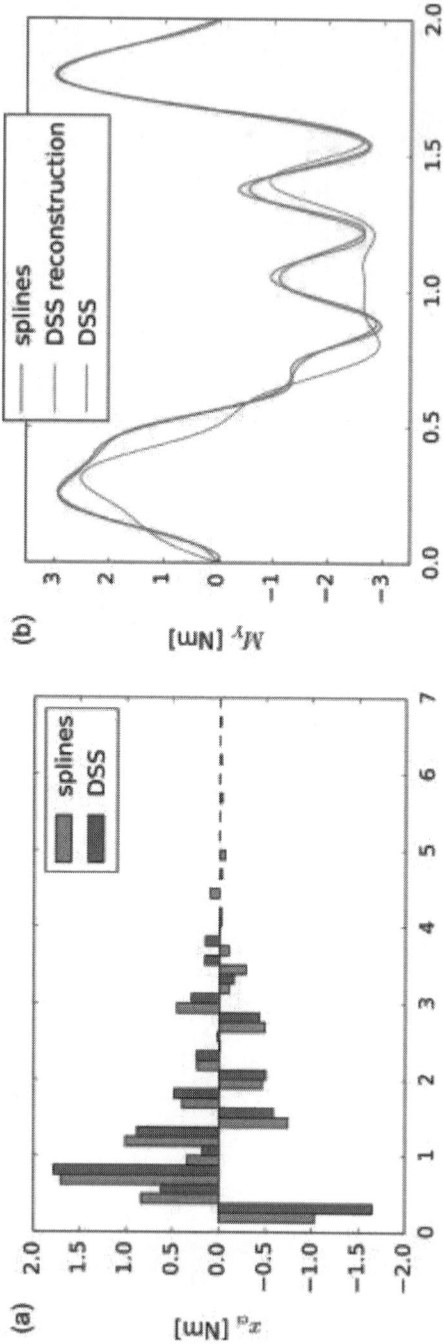

(a) DSs comparison between optimal actuations

(b) Torque comparison and reconstruction of spline optimal actuation via DSS

Figure. 10. Comparison between spline and DSS optimal actuation for *flexible* pendulum case and no penalty factor.

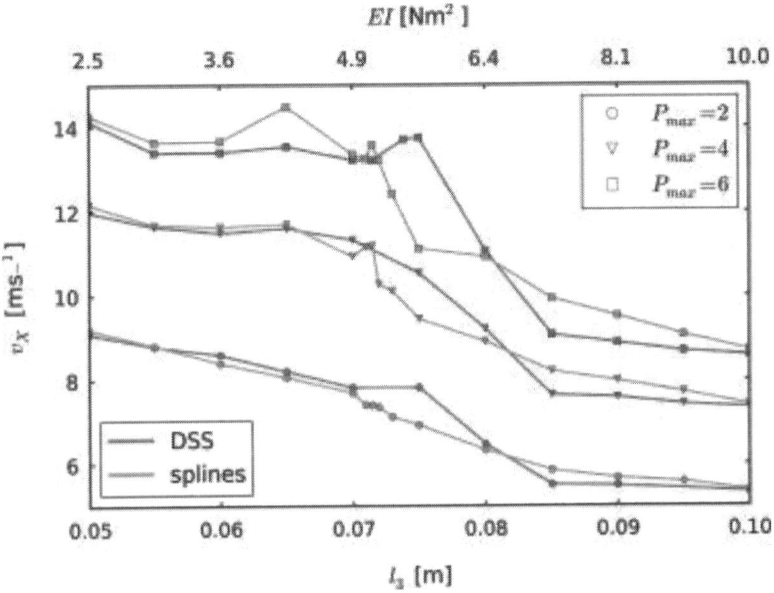

Figure. 11. Final tip velocity, $vX(T\ \ )$, obtained using different parametrisations and levels of actuation P_{max}.

This is finally demonstrated in Fig. 13, where the optimal actuations provided by the two parametrisations for the case l_3=0.0750 m and P_{max}=4 are compared. The DSS representation of the two signals (Fig. 13a) shows that the DSS parametrised control has a higher frequency content around the bending natural frequency, $f_b = 5.63\,\text{Hz}$: this allows to better excite the mode and achieve a larger final tip velocity, $vX(T)$. As observed in Fig. 13b, however, the optimal actuation obtained using a DSS cannot be reproduced using B-splines. As a result, in the transition region DSS outperforms B-splines as it can better excite higher frequencies.

3.4. A multi-resolution strategy

Finally, a multi-resolution approach is proposed to improve optimisation convergence when strong structural nonlinearities are present. Fig. 11 has shown that for high frequency actuations and mild level of nonlinearity (P_{max}=2,4), results compare well across different parametrisations. Both DSS and B-splines capture accurately the relevant high frequencies necessary to induce the system resonance. When the level of actuation is increased (P_{max}=6), peaks of resonance become more evident as a consequence of larger displacements. These tend to arise whenever the bending oscillation period is a divisor of the time horizon *T*. While the location of the peaks is well captured by both basis, their amplitude can become parametrisation dependent. In fact, around the peaks of resonance there are large changes in performance for small variations in the actuation. There local basis tend to perform better, as they scale more efficiently. During the sensitivity analysis, in fact, a variation applied to one of the basis functions only affects a portion of the domain, leading to an overall smaller perturbation of the original signal. For the same reason, when updating the actuation, changes in torque signal tend to be more contained than when using a DSS. As a consequence, smaller steps between one design point and another are taken by the optimiser. This facilitates the reconstruction of the design space around the peak regions, despite the high sensitivity of the cost function to variations of the design.

A multi-resolution strategy can be used to accelerate convergence on those situations. The optimal control problem is hierarchically solved for parametrisations of increasingly higher bandwidth. The method is demonstrated here for the case of a pendulum of bending stiffness EI=6.4 N m^2 and whose actuation is obtained through a large bandwidth DSS parametrisation $(f_{max} = 12\text{ Hz})$. The steps of the process are summarised in Table 5, where the results obtained without applying a multi-resolution strategy are also reported (*direct* label). Initially, the actuation bandwidth is set to 6 Hz. Once an optimal actuation is found, this is increased by 2 Hz so as to refine the torque time history; the procedure is repeated until $f_{max} = 12\text{ Hz}$.

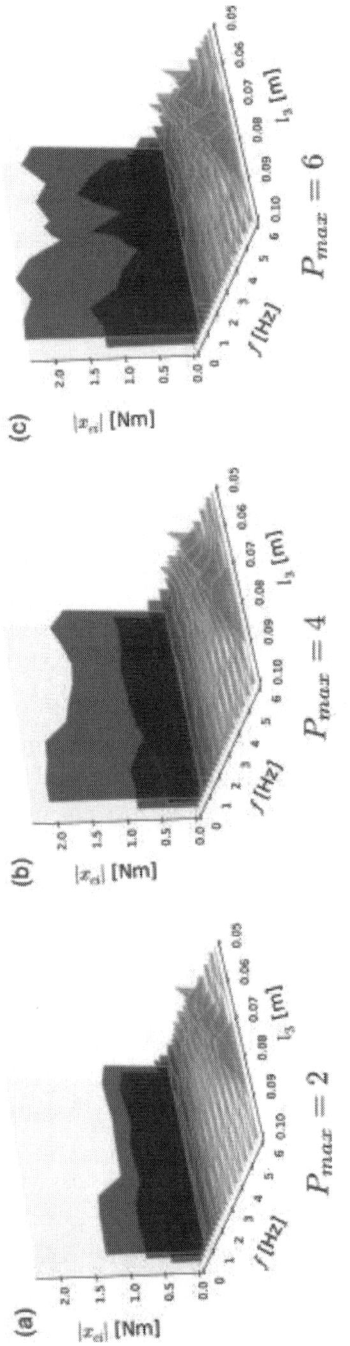

Figure. 12. Optimal actuation frequency content ($|x_{ci}|$) using a DSS parametrisation ($f_{max} = 6$ Hz) and a constrained problem formulation (13). Different levels of actuation, P_{max}, are allowed.

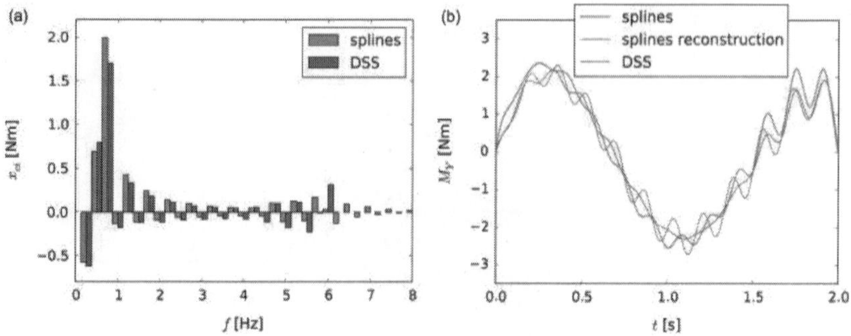

DSS optimal coefficient against DS representation of spline solution

Time histories and spline reconstruction of DSS optimal actuation

Figure. 13. Optimal actuation in the transition region (l_3=0.0750 m and P_{max}=4) using different parametrisations.

As expected, the pendulum final tip velocity, $vX(T\ \ )$, is drastically increased at the second stage of the process, as the actuation bandwidth becomes larger than the pendulum bending mode natural frequency $(f_b = 6.21\ \mathrm{Hz})$. Further increments to the maximum actuation frequency, f_{max}, however, do not lead to any relevant change in $vX(T\ \ )$: contrarily, as the parametrisation becomes over-resolved ($f_{max} \gg f_b$) the convergence of the optimisation process is compromised and the constraint $P \leq 6$ is not fully recovered. The link between over-resolution and convergence slowdown is even more clear when looking at the results obtained without multi-resolution analysis. In this case, not only the constraint $P \leq 6$ is not fully recovered, but also the final actuation time history shows to be not fully converged, as $vX(T)$ achieved is 8.9 percent lower than the one provided by the multi-resolution approach.

4. NUMERICAL STUDIES ON ACTIVE SYSTEM CO-DESIGN

The same flexible pendulum test case of Section 3 is used now to test the co-design approach presented in Section 2.2. As the pendulum bending stiffness has been shown to be the most relevant structural property to impact the system dynamics, it will be allowed to vary while the actuation itself is being defined. The effect of choosing a different starting condition for the actuation, as well as the potential disadvantages of using a sequential approach, is investigated in Section 4.1. Section 4.2 focuses on the efficiency of the process, showing how the convergence speed can be affected near resonance and how to tackle the issue. Finally, the impact of the frequency resolution chosen to model the actuation is discussed in Section 4.3.

For all the co-design cases considered in this section, the beam rectangular cross-section (Fig. 2) is initially set to have a relatively high bending stiffness, EI=6.4 N m^2, and size $l_2 \times l_3 = 0.080 \times 0.125\ \mathrm{m}^2$. During the co-design, l_3 has been bounded to be $l_3 \in [0.055\ \mathrm{m}, 0.200\ \mathrm{m}]$, while l_2 is constrained as before to maintain the beam cross-sectional area constant $(A = 10^{-2}\ \mathrm{m}^2)$. As the bending mode natural frequency of the initial design is $f_b = 6.21\ \mathrm{Hz}$, the basis size of the parametrisation has been adjusted so as to reach a maximum frequency f_{max} equal to $4\ \mathrm{Hz}$, $6\ \mathrm{Hz}$ (as per the design of experiment inFig. 11) and $12\ \mathrm{Hz}$. The optimisation has been formulated as the constrained problem (13)with P_{max}=6. This allows to test the co-design process for a highly nonlinear configuration while reducing the path dependency of the solution.

The constraint on the sectional area maintains the inertial properties and thus the characteristic frequencies associated to the rigid-body dynamics of the different designs remain identical. To minimise the cost in problem (13), therefore, the co-design can only adjust the pendulum stiffness and tune its bending mode natural frequency with the frequencies content of the actuation. While a larger number of structural design parameters could be considered, a single parameter facilitates our exploration as it allows for a much clearer analysis of the results. Furthermore, dealing with nonlinear dynamics and a large number of structural design variables would require the development of an adjoint based sensitivity analysis, which is beyond the scope of this work.

Table 5. Optimal control results at different stages of a multi-resolution based process using a DSS parametrised actuation ($f_{max} = 12$ Hz).

Stage	N_c	f_{max} (Hz)	$v_x(T)$ (m s^{-1})	P
1	24	6	10.71	6.004
2	32	8	13.41	6.004
3	40	10	13.51	6.038
4	48	12	13.54	6.051
direct	48	12	12.22	6.019

The main results obtained of the co-design problem have been summarised inTable 6 and Table 7. For each case, the size of the parametrisation, *Nc*, and the related maximum frequency of excitation, f_{max}, have been reported. The identifications *zero*and *opt* refer to the torque time-history at the beginning of the optimisation process, MY_0: For the *opt* cases, the initial design uses the actuation found in pure optimal control problem; for the *zero* cases, it is simply chosen to be $M_{Y0}(t)=0$ (*zero* cases). The optimal structural design is described both in terms of final geometry (l_3) and its bending mode natural frequency (f_b), while system performance is quantified by the final tip velocity, $vX(T)$. This is compared to the tip velocity given by the optimal control problem on the initial beam geometry (DOE in the tables).

4.1. Initial condition: sequential vs. co-design

First, we will focus on the co-design results obtained when initialising the torque actuation with a previously computed pure optimal control solution (*opt* cases). This defines a situation similar to a sequential optimisation, and in fact, fromTable 6 and Table 7, it is clear that co-design brings almost no changes in either the structural design or in the actuation with respect to the initial design. This can also be observed by comparing the *DOE* and *opt* curves in Fig. 14, where the optimal actuations obtained with 4 Hz and 6 Hz bandwidth parametrisations are shown.

For the large bandwidth controls ($f_{max} = 6\,\mathrm{Hz}$, 12 Hz) this can be explained noticing that at the start of the optimisation the actuation MY_0, which is the solution of an optimal control problem, can already excite the bending mode of the pendulum (note, for instance, the high frequency component in the *DOE* curve in Fig. 14a). As the process is initiated from a resonance condition, control force and the structural design are de facto *locked* into a local minimum. Further changes in the structure size would only lead to poorer performance, the system moving away from the resonance condition. Conversely, changes in the actuation signal would also imply moving away from resonance. As no descending direction can be found, the optimisation is interrupted after few iterations.

When the starting torque includes only lower frequencies ($f_{max} = 4$ Hz), the phenomenology that limits the performance improvements is similar. The initial excitation (see the *DOE* curve in Fig. 14b) exploits the rigid-body dynamics of the pendulum and, being far from resonance conditions, system performance can initially only be improved by stiffening the design, which ensures that all the energy transferred to the pendulum is stored by the rigid mode. Consequently, the co-design leads to a stiffer beam with only a small refinement on the actuation so as to better exploit the rigid-body dynamics of the system (see the *opt* curve in Fig. 14b).

The changes in the control design space leading to *locking* can be better understood with the reduced basis already introduced in Section 3.2.1. To reproduce a high-bandwidth actuation, the amplitude of the sine basis of frequencies 0.5 Hz ($x_{0.5}$) and 7.75 Hz ($x_{7.75}$) was varied, while the 0.75 Hz wave amplitude was fixed to 1.7 N m. The final tip velocity of the pendulum, $vX(T\ )$, associated to different points ($x_{0.5}$,$x_{7.75}$) is shown inFig. 15a. The isolines of the force regulator term P are also shown (black curves). Three high velocity regions, strongly dependent on the amplitude of the 7.75 Hz wave, are observed. For P≤6, the maximum velocity, $v_X(T) = 9.36\ \mathrm{m\ s^{-1}}$, is found at point A =(0.95,−0.49) N m. This is associated to a condition of resonance, as the bending mode natural frequency is $f_b = 7.76$ Hz. As the initial structural design (l_3=0.1 m) is varied, the system moves away from the resonance. The high velocity regions initially move (Fig. 15b and c) and gradually smooth out (Fig. 15d and e). This shows that in the co-design space (l_3,$x_{0.5}$,$x_{7.75}$) a maximum is located around l_3=0.1 m, i.e. precisely where the bending mode natural frequency tunes with the actuation frequency of 7.75 Hz.

Table 6. Results of the combined optimisation problem using DSS control parametrisations of different basis size. Percentage values are with respect to the optimal control only case (DOE).

ID	N_c	$f_{\max}$ (Hz)	l_3 (m)	f_b (Hz)	$v_X(T)$ (m s^{-1})	%
zero	16	4	0.0550	4.27	10.61	33.5
zero	24	6	0.0739	5.73	13.70	28.0
zero	48	12	0.0792	6.15	13.34	−0.5
opt	16	4	0.0819	6.36	8.41	5.8
opt	24	6	0.0800	6.21	10.71	0.0
opt	48	12	0.0800	6.21	13.41	0.0
DOE	16	4	0.0800	6.21	7.95	–
DOE	24	6	0.0800	6.21	10.71	–
DOE	48	12	0.0800	6.21	13.41	–

Table 7. Results of the combined optimisation problem using spline control parametrisations of different basis size. Percentage values are with respect to the optimal control only case (DOE).

ID	N_c	f_{max} (Hz)	l_3 (m)	f_b (Hz)	$v_X(T)$ (m s^{-1})	%
zero	19	4	0.0550	4.27	12.55	45.6
zero	27	6	0.0717	5.56	13.54	24.0
zero	51	12	0.0828	6.43	12.64	−9.2
opt	19	4	0.1465	11.4	8.65	0.3
opt	27	6	0.0800	6.21	10.92	0.0
opt	51	12	0.0800	6.21	13.71	0.0
DOE	19	4	0.0800	6.21	8.62	–
DOE	27	6	0.0800	6.21	10.92	–
DOE	51	12	0.0800	6.21	13.71	–

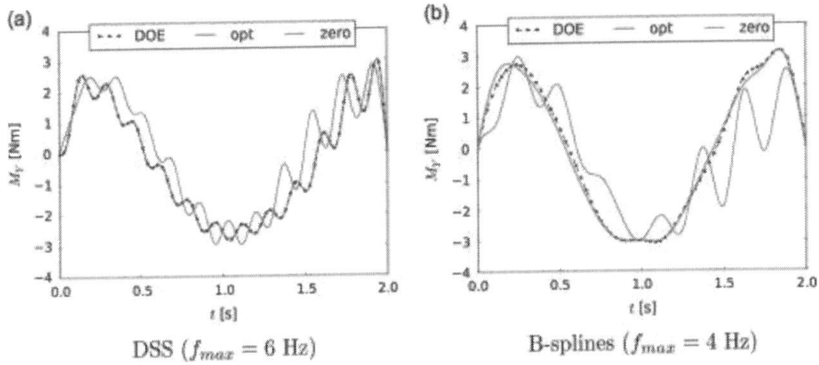

Figure. 14. Optimal actuation provided by the co-design using different starting conditions and parametrisations.

The mechanism determining the position of the high tip velocities regions in Fig. 15a is linked to the phase between the high frequency component of the actuation and the response of the pendulum in terms of its tip horizontal position. The global maximum (point A) is associated to the rigid-flexible body dynamics, while a second minimum, associated to predominantly rigid-body dynamics, is located at point B=(1.00,0.063) N m. The actuation torque, *MY*, and the *X* tip position time histories of these cases are shown in Fig. 16 for t∈[1.8,2] s. As expected, for the actuation corresponding to point A, the high frequency component of *MY* synchronises with the displacements in such a way that, as t→T, the tip velocity due to the structural vibration adds to the rigid-body rotation. Note that this is achieved for a negative amplitude of the 7.75 Hz wave, $x_{7.75}$. As expected, when $x_{7.75}$ increases and changes sign, the final tip velocity reduces because resonance provides a negative velocity increment. When, however, $x_{7.75}\gtrsim 0.5$ N m, $vX(T)$ can be observed to grow again (Fig. 15a), meaning that the phase between actuation and pendulum tip response also depends on the actuation amplitude,$x_{7.75}$. This is also shown in Fig. 16, where the time histories of the actuation corresponding point C=(1.10,0.98) N m in Fig. 15a are presented. This nonlinear effect is linked to the large rigid-body rotations of the pendulum and the changing geometric stiffness.

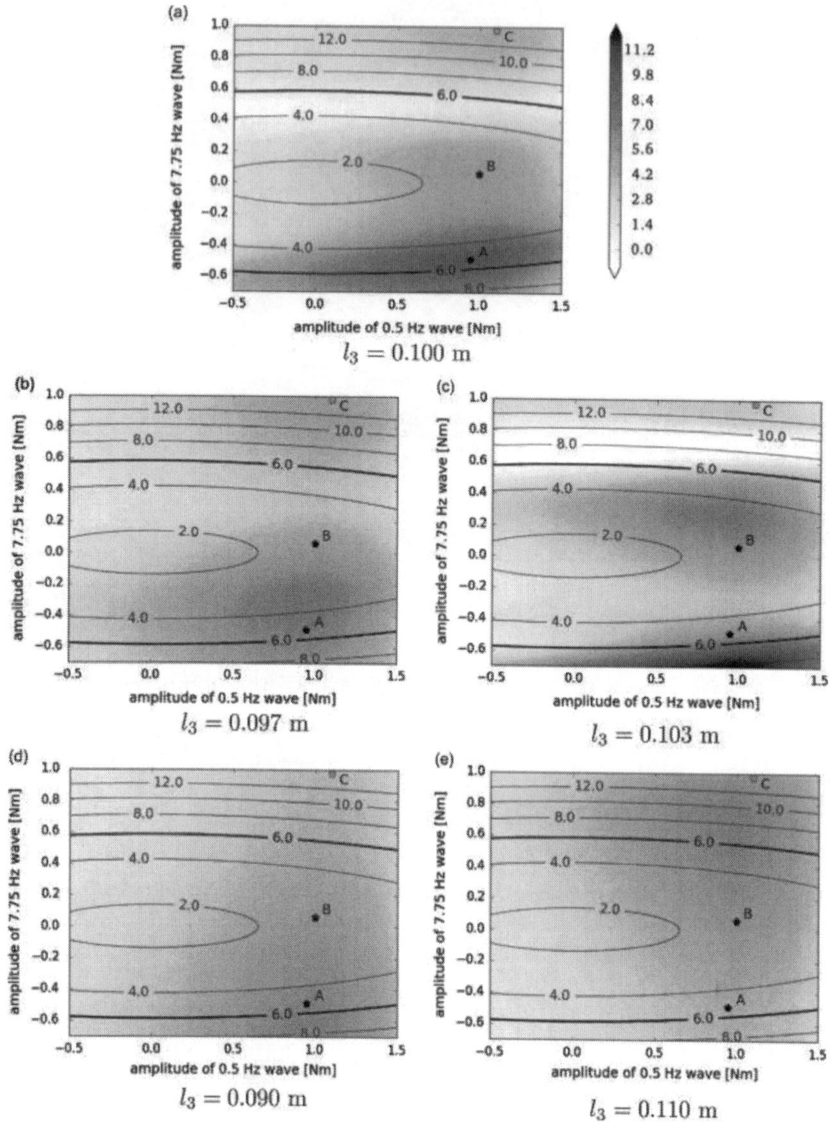

Figure. 15. Visualisation of velocity profile for a high bandwidth actuation, modelled with a reduced number of sine waves, for structural designs in the neighbourhood of $l_3 = 0.1$ m. Points *A*, *B* and *C* correspond to the coordinates of the local maxima found for $l_3 = 0.1$ m.

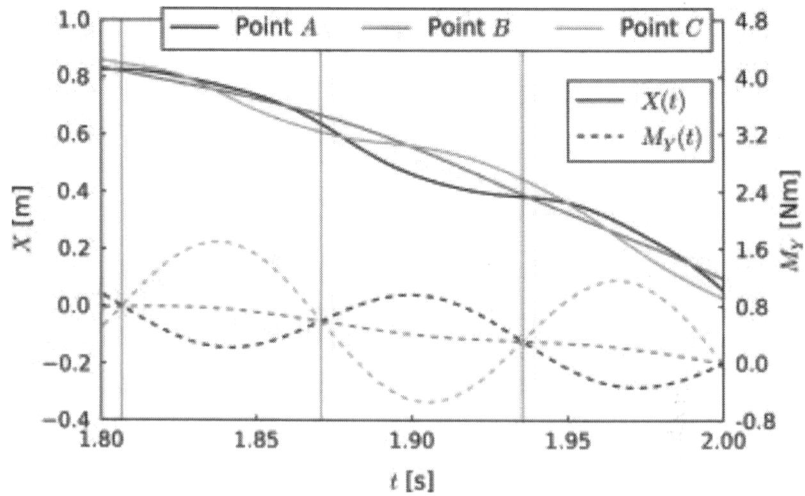

Figure. 16. Tip horizontal position and torque moment *MY* time histories of the *stiff* pendulum under the reduced DSS parametrised actuations corresponding to points A, B and C in Fig. 15a.

4.2. Robustness of the design

Contrarily to what it was observed in the previous section, when a null starting torque is used (*zero* cases in Table 6 and Table 7), structural design and final system performance undergo a significant evolution. The outcome of the process largely depends in this case mostly on the control bandwidth. Generally, the co-design is driven towards a peak of resonance region, which, as it was already seen in Section 3.3, implies a slowdown of convergence, particularly for DSS parametrised controls. As a result, the optimiser typically does not recover the integral constraint $P \leq P_{max}$. To overcome this, the following two-stage strategy was found to be an effective solution: once variations in the structural design are negligible, the pendulum geometry is fixed (end of first stage) and the actuation is refined by solving a pure optimal control problem (second stage). The improvements obtained by using the two-stage approach to refine the actuation are summarised in Table 8; as expected, the benefit is higher when the solution is closer to a peak of resonance.

Table 8. Performance gains obtained using a two-stage co-design strategy.

Basis	f_{max} (Hz)	l_3 (m)	$v_X(T)$ (m s^{-1})		Gain %
			Not refined	Refined	
splines	4	0.0550	12.55	12.55	0.0
splines	6	0.0717	11.55	13.54	17.2
splines	12	0.0828	12.35	12.64	2.3
DSS	4	0.0550	10.50	10.61	1.0
DSS	6	0.0739	12.56	13.70	9.1
DSS	12	0.0792	11.72	13.34	13.8

The causes of the convergence slow down can be further investigated referring to the spline parametrised solution obtained with a 6 Hz bandwidth control. In this case, the optimiser correctly drives the structural design towards the peak region located around $l_3 = l_{ref} = 0.0715$ m (curve P_{max}=6 in Fig. 11). The neighbourhood of the peak of resonance is shown in more detail in Fig. 17. Here the coefficients of the optimal actuation x_c (Fig. 17b), and the resulting control signal (Fig. 17d), are compared for structural designs in the vicinity of the peak of resonance. Despite the variations in the structure size being small,[3] $\Delta l_3 = \pm 0.007 l_{ref}$, the changes in the optimal actuation are, in proportion, considerably larger. This can be easily observed by comparing Fig. 17b against Fig. 17a, where the optimal control coefficients around $l_3 = l_{ref} = 0.0715$ m, found enforcing a lower cost of actuation (P_{max}=2), are shown. Note that, in the latter case, nonlinear effects are significantly smaller.

The change in optimal actuation can be quantified by measuring the integral norm

$$\epsilon = \frac{\int_0^T (M_Y - M_{Yref})^2 \, dt}{\int_0^T M_{Yref}^2 \, dt} \tag{14}$$

where MY_{ref} and MY are the optimal actuations at the reference, $l_{ref} = 0.0715$ m, and the perturbed, $l_3 = l_{ref} \pm \Delta l_3$, design points (Fig. 17d). While perturbations in structural design are around 0.7 percent, the norm in Eq. (14) reaches values 16 times higher, being in both cases $\epsilon \approx 11$ percent. Large values of ϵ imply that, fixing the actuation and applying even a small change in the structural design, the system will tend to move far away from the optimal control condition. Not only, therefore, performance will decrease consistently, but, as discussed in Section 3.3, the parametrisation of the actuation will usually require a large number of iterations to adjust to the new structural design.

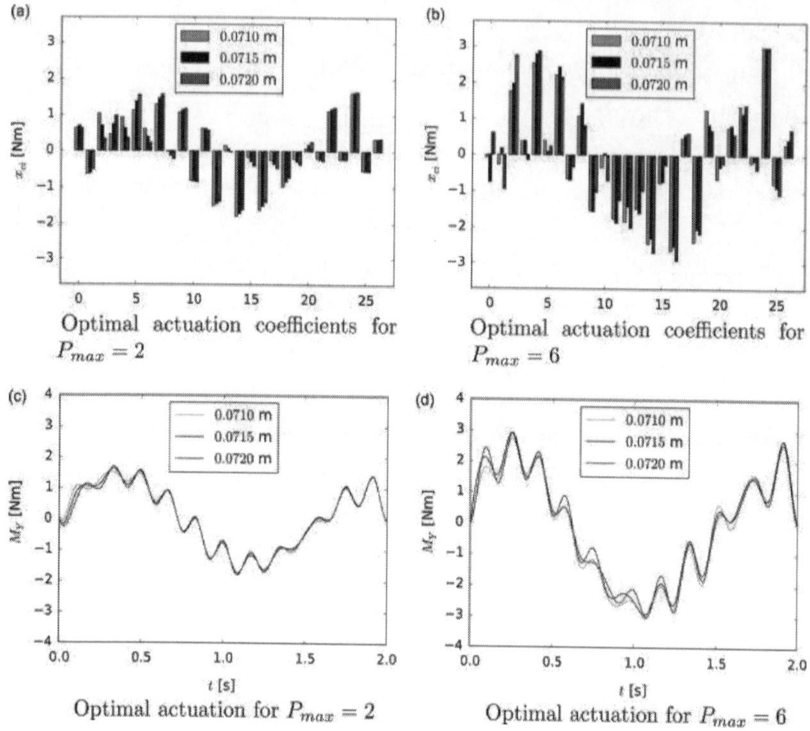

Figure. 17. Optimal control torque (parametrised with B-splines) for structural designs in the proximity of $l_{ref} = 0.0715$ m for small (P_{max}=2) and large (P_{max}=6) amplitude oscillations.

4.3. Impact of frequency resolution

From Table 6 and Table 7, it can be observed that setting $M_{Y0}(t)=0$ at the beginning of the co-design generally gives better performance, and that improvements are consistently larger when reducing the bandwidth of the control. When the highest bandwidth control is used ($f_{max} = 12$ Hz), the actuation is capable of inducing resonance also on the initial design. Consequently, no major changes in the structural design occur, as this is driven towards the nearest peaks of resonance. The disadvantage of exploring a non-smooth region using a gradient based approach is shown by the fact that, while the DS parametrisation improves performance, with B-splines it degenerates and the solution converges towards a less resonant region.

Decreasing the actuation bandwidth, instead, leads to important performance increases with respect to the optimal control problem, regardless of the parametrisation used. The pendulum stiffness is, in fact, always reduced so as to drive its bending mode natural frequency towards the range on actuation of the control (note the f_b values inTable 6 and Table 7). When $f_{max} = 6\text{ Hz}$, the structural design is moved towards the peak of resonance located in the region $l_3 \in (0.070\text{ m}, 0.075\text{ m})$ and clearly visible from the design of experiment in Fig. 11. When $f_{max} = 4\text{ Hz}$, the increase in flexibility is even larger and the lower bound of the structural design space, l_3=0.0550 m, is hit.

Especially in this case, the combined optimisation is shown to be capable of driving the structural design to a configuration where the initial control (with its limitations in terms of maximum frequency resolution) can now fully exploit the physical properties of the system. The performance gains with respect to the optimal control only case (DOE label in Table 6 and Table 7) are as large as 45 percent, due to the switch from rigid to flexible-rigid body dynamics. This effect is clearly reflected by the development of a high frequency component in the optimal actuation signal (*zero* label in Fig. 14b). Also for the $f_{max} = 6\text{ Hz}$ case, where a mild level of resonance can already be induced on the initial design, the stiffness reduction achieved via co-design allows to better exploit it (compare the *DOE* and *zero* curves in Fig. 14a).

The dependency of the co-design derived optimal solution on the actuation bandwidth has important implications when a multi-resolution strategy (Section 3.4) is adopted to solve the co-design problem. When the combined optimisation is hierarchically solved using actuations of increasing bandwidth, in fact, the final structural design is driven by the bandwidth used to initialise the process. This is shown in more detail in Table 9, where the integrated multi-resolution design approach is demonstrated for a case in which the actuation is parametrised with a DSS. The optimisation is initiated using both a control bandwidth of 4 Hz and 6 Hz (stage 0): once the first combined optimisation is terminated, this is restarted after incrementing the maximum actuation frequency, f_{max}, of 2 Hz. Note that at stage 0 structural design and actuation are as per the 4 Hz and 6 Hz bandwidth *zero* cases in Table 6.

When the multi-resolution process is started with a 6 Hz bandwidth actuation, *locking*between structural design and control force prevents any changes of the structural properties. Therefore, when a 12 Hz bandwidth is reached, the integrated design provided by the 12 Hz *zero* case in Table 6 is not retrieved. The final tip velocity value,$vX(T)$, is, however, comparable as in both cases resonance is exploited. When the process is started from a 4 Hz bandwidth solution, on the other hand, structural design and actuation are not initially *locked* in a peak of resonance, being the bending natural frequency of the pendulum (fb=4.27) outside the actuation range. As a result, design and actuation are refined during the first stage and the final tip velocity of the pendulum,$vX(T)$, increases by 7.1 percent. Once, however, structural design and actuation are*locked*, no further change is observed as the actuation bandwidth is increased.

The solutions in the multi-resolution process obtained from an initial 4 Hz bandwidth actuation in Table 9 can be compared against the corresponding co-design solutions inTable 6. In all cases, the final tip velocities and the level of actuation are very similar. However, the structural design obtained via multi-resolution is consistently more flexible, showing that the initial actuation bandwidth has a clear impact on the final structural design.

Overall, the results obtained using a zero starting condition for the actuation prove the potential of adopting a gradient based co-design approach even for resonant, strongly nonlinear systems. On the one hand, and similar to what was observed with a sequential optimisation strategy (Section 4.1), the occurrence of structural-actuation *locking* around peaks of resonance can limit the progress of the optimisation, especially when a large bandwidth is used. This means that some initial sampling of the structural domain needs to be performed. On the other hand, however, co-design allows to explore much larger portions of the design space, even when this implies a drastic change in the system dynamics (such as those switching from a rigid to a flexible-rigid body dynamics). The rate of sampling of the structural design space, or conversely the overall number of co-design solutions to be found, can, thus, be conveniently reduced, facilitating the exploration of larger portions of the design space as compared to a zero order or traditional sequential optimisation approaches.

5. CONCLUSIONS

The control vector parametrisation technique has been used to solve a nonlinear optimal control problem, defined on a very flexible structure, and to explore the co-design process of open-loop control and structural properties of the system. The problem has been chosen to represent key features of the control of very flexible structures and a gradient method has been used to drive the optimisation. The control has been modelled using B-splines and DSS to test the effect of using local and global sets of basis functions. In both the optimal control and the co-design cases the time–frequency resolution of the parametrisations has been shown to be the most relevant factor driving the design.

In the optimal control problems, both the effect of an increasing rigid-flexible body dynamics coupling and the impact of the problem definition on the optimisation outcome have been investigated. The process has been demonstrated to capture and exploit the structural resonances, regardless of the level of nonlinearity shown by the system. The actuation cost has been measured by means of a quadratic force regulator term and included in the problem firstly via penalty augmentation and then via constraint enforcement. In the first case, a dependency of the optimal control on the path taken by the optimiser — which increases with the system nonlinearity — has been found. The constraint enforcement approach has allowed, instead, to produce more consistent results across parametrisations, as it drives the optimal control to reach the maximum admissible level of actuation.

An assessment of the impact of the parametrisation on the optimal control solution process has also been carried out. A transition region, in which the different reconstruction properties of splines and DSS leads to a gap in the final achievable performance, was identified. Elsewhere a good consistency across parametrisations has been found, particularly for low-medium levels of actuation and, thus, moderate nonlinear effects, while increasing the authority of the control leads to multiple local peaks of resonance. In this last case, the design space was found to be particularly non-smooth, causing the optimisation process to converge slowly, particularly when using global basis. A multi-resolution strategy has been proposed in these cases to make the process more efficient.

In the co-design problem, the system flexibility was allowed to vary while defining the actuation law itself. We have shown that co-design attempts starting from a previously computed optimal control solution (which are equivalent to a sequential design approach) do not result in relevant changes to structure size and control law. Once tuned on the same resonance/excitation frequency, structure and control *lock* in the initial design.

Designing the control law from zero has the great advantage of avoiding *locking* and, thus, allowing to move through larger portions of the design space. The high level of nonlinear effects nearby resonance regions has been shown to affect the convergence speed of the optimisation. The phenomenon, connected to the increased sensitivity of the control law with respect to changes in the structural design, requires a two-stage strategy — via solution of a pure optimal control problem — to fully complete the process.

Table 9. Results of the combined optimisation using a multi-resolution strategy and a DSS parametrisation. Two different actuation bandwidth of 4 Hz and 6 Hz has been chosen to start the process.

Case	Stage	N_c	f_{max} (Hz)	l_3 (m)	f_b (Hz)	$v_X(T)$ (m s^{-1})
Initial bandwidth of 4 Hz	0	16	4	0.0550	4.27	10.61
	1	24	6	0.0584	4.53	13.44
	4	48	12	0.0584	4.53	13.44
Initial bandwidth of 6 Hz	0	24	6	0.0739	5.73	13.70
	3	48	12	0.0739	5.73	13.70

The key element in the co-design process is the actuation time–frequency resolution. When the control bandwidth is larger than the structure bending mode natural frequency, the gradient based optimiser drives, as expected, the structural design towards the nearest peaks of resonance. However, when using a lower bandwidth control, performance is drastically improved with both parametrisations. Larger portions of the design space can be explored and the combined optimisation has always managed to successfully drive the structural design in a region where the control could exploit the coupled rigid-flexible dynamics. While the portability of the approach when dealing with more complex structural systems — e.g. whose dynamics is driven by more modes of vibration — still requires a further assessment, these results prove the potential of gradient based co-design in vibrations control and structural dynamics, even for highly nonlinear problems.

ACKNOWLEDGEMENTS

The authors acknowledge the support of the UK Engineering and Physical Science Research Council (EPSRC), including an access grant to ARCHER, the UK National Supercomputing Service. All results in this paper are available as open data. Information on access can be found at http://www.imperial.ac.uk/aeroelastics/software.

REFERENCES

1. J. Allison, D.R. Herber, Multidisciplinary design optimization of dynamic engineering systems, AIAA Journal 52 (4) (2014) 691–710, http://dx.doi.org/ 10.2514/1.J052182.

2. H. Fathy, J. Reyer, P. Papalambros, A. Ulsoy, On the coupling between the plant and controller optimization problems,

Proceedings of the American Control Conference, Arlington, 2001, pp. 1864–1869.

3. J.R.R.A. Martins, A.B. Lambe, Multidisciplinary design optimization: a survey of architectures, AIAA Journal 51 (9) (2013) 2049–2075, http://dx.doi.org/ 10.2514/1.J051895.

4. J. Onoda, R. Haftka, An approach to structure/control simultaneous optimization for large flexible spacecraft, AIAA Journal 25 (8) (1987) 1133–1138.

5. S.S. Rao, Combined structural and control optimization of flexible structures, Engineering Optimization 13 (1) (1988) 1–16, http://dx.doi.org/10.1080/ 03052158808940943.

6. H. Asada, J. Park, S. Rai, A control-configured flexible arm: integrated structure control design,, Proceedings of the 1991 IEEE International Conference on Robotics and Automation, Sacramento, California, 1991, pp. 2356–2362,

7. Z. Xianmin, L. Jianwei, S. Yunwen, Simultaneous optimal structure and control design of flexible linkage mechanism for noise attenuation, Journal of Sound and Vibration 299 (4–5) (2007) 1124–1133

8. N. Alujević, G. Zhao, B. Depraetere, P. Sas, B. Pluymers, W. Desmet, H2 optimal vibration control using inertial actuators and a comparison with tuned mass dampers, Journal of Sound and Vibration 333 (18) (2014) 4073–4083,

9. H. Nishigaki, K. Kawashima, Motion control and shape optimization of a suitlike flexible arm, Structural Optimization 15 (1998) 163–171.

10. J.B. Cardoso, P.P. Moita, A.J. Valido, Design and control of nonlinear mechanical systems for minimum time, Shock and Vibration 15 (3–4) (2008) 315–323

11. J.T. Allison, T. Guo, Z. Han, Co-design of an active suspension using simultaneous dynamic optimization, Journal of Mechanical Design 136 (8) (2014) 14

12. S. Haghighat, J.R.R.A. Martins, H.H.T. Liu, Aeroservoelastic design optimization of a flexible wing, Journal of Aircraft 49 (2) (2012) 432–443

13. T. Jackson, E. Livne, Integrated aeroservoelastic design optimization of actively controlled strain-actuated flight vehicles, AIAA Journal 52 (6) (2014) 1105–1123

14. D. Trivedi, D. Dienno, C. Rahn, Optimal, model-based design of soft robotic manipulators, Journal of Mechanical Design 130 (9) (2008) 091402

15. V. Etxebarria, A. Sanz, I. Lizarraga, Control of a lightweight flexible robotic arm using sliding modes, International Journal of Advanced Robotic Systems 2 (2) (2005) 103–110.

16. S. Dwivedy, P. Eberhard, Dynamic analysis of flexible manipulators, a literature review, Mechanism and Machine Theory 41 (7) (2006) 749–777, http:

17. C.E. Cesnik, R. Palacios, E.Y. Reichenbach, Reexamined structural design procedures for very flexible aircraft, Journal of Aircraft 51 (5) (2014) 1580–1591

18. T. Önsay, A. Akay, Vibration reduction of a flexible arm by time-optimal open-loop control, Journal of Sound and Vibration 147 (2) (1991) 283–300

19. L.T. Biegler, Nonlinear Programming, Concept, Algorithms, and Applications to Chemical Processes, SIAM, Philadelphia, 2010.

20. M. Schlegel, K. Stockmann, T. Binder, W. Marquardt, Dynamic optimization using adaptive control vector parameterization, Computers & Chemical Engineering 29 (8) (2005) 1731–1751

21. B.C. Fabien, Piecewise polynomial control parameterization in the direct solution of optimal control problems, Journal of Dynamic Systems, Measurement, and Control 135 (3) (2013).

22. Q. Lin, R. Loxton, K.L. Teo, The control parameterization method for nonlinear optimal control: a survey, Journal of Industrial and Management Optimization 10 (1) (2013) 275–309

23. D.H. Hodges, A. Mixed, Variational formulation based on exact intrinsic equations for dynamics of moving beams, International Journal of Solids and Structures 26 (11) (1990) 1253–1273

24. M. Geradin, A. Cardona, Flexible Multibody Dynamics: A Finite Element Approach, John Wiley & Sons Ltd, Chichester, UK, 2001.

25. H. Hesse, R. Palacios, Reduced-order aeroelastic models for dynamics of maneuvering flexible aircraft, AIAA Journal 52 (2014) 1–16

26. R. Palacios, Nonlinear normal modes in an intrinsic theory of anisotropic beams, Journal of Sound and Vibration 330 (8) (2011) 1772–1792

27. C. de Silva, Vibration and Shock Handbook, CRC Press, Boca Raton, Florida, USA, 2005.

28. D. Kraft, A Software Package for Sequential Quadratic Programming, Technical Report, DLR German Aerospace Center Institute for Flight Mechanics, Köln, Germany, 1988.

29. J. Murua, R. Palacios, J.M.R. Graham, Applications of the unsteady vortex-lattice method in aircraft aeroelasticity and flight dynamics, Progress in Aerospace Sciences 55 (2012) 46–72

30. J.S. Gray, K.T. Moore, T.A. Hearn, B. Naylor, Standard platform for benchmarking multidisciplinary design analysis and optimization architectures, AIAA Journal 51 (10) (2013) 2380–2394

31. Q. Wang, W. Yu, Sensitivity analysis of geometrically exact beam theory (GEBT) using the adjoint method with hydra, The 52nd AIAA/ASME/ASCE/AHS/ ASC Structures, Structural Dynamics and Materials Conference, Denver, Colorado, 2011, pp. 1–15.

32. M. Sabatini, P. Gasbarri, G.B. Palmerini, R. Monti, Operational modal analysis via image based technique of very flexible space structures, Acta Astronautica 89 (2013) 139–148

CHAPTER 2

Vibration Control of a Gym Floor Using Tuned Mass Dampers: A Numerical Analysis

Marcos Daniel Souza dos Santos, Daniel Valença Fiuza Lima, Jorge Eliécer Campuzano Carmona, Suzana Moreira Avila, Graciela Nora Doz de Carvalho

Departamento de Engenharia Civil e Ambiental, Faculdade de Tecnologia, Universidade de Brasilia, Brasilia, Brasil

ABSTRACT

This work presents a study on excessive vibration problem occurring on concrete slabs, usually used on residential and commercial building floors. Even well designed slabs, according to ultimate and serviceability limit states criteria, can be vulnerable to undesirable vibrations that lead to user discomfort. A gym floor, that presented real excessive vibrations, located in a commercial building situated in the city of Brasilia, Brazil, was analyzed via Finite Element Method using ANSYS software. The first step in this analysis was to obtain natural frequencies and vibration modes, the structure presented low natural frequencies representing its flexible behavior. Then it was simulated a dynamic loading of people jumping, characteristic of this type of building occupation. Since it was observed the occurrence of excessive vibrations also in the numerical analysis, a Tuned Mass Damper (TMD) control system was proposed, looking for the best set of dampers to improve the control performance. The parameters for the best vibration reduction were obtained via a parametric study considering four different slabs varying dimensions and support conditions. Different models considering one and more TMDs, varying its placements and parameters, besides the frequency reference value to tune the damper were considered. An efficient control solution to this practical problem is presented to reduce its undesirable vibrations.

KEYWORDS

Structural Dynamics; Structural Control; Concrete Slabs; Tuned Mass Damper

1. INTRODUCTION

Human induced loads can cause excessive vibration leading to discomfort and even compromising structural safety. Real occurrences of this problem in structures subjected to loads originated from people movement are worldwide known, a famous example is the London Millenium Footbridge [1]. Floor vibration studies have intensified in the last years. A number of factors cause this effect; the more important are: the use of more resistant materials that lead to more slender and flexible slabs, and the occurrence of non-provided activities in the original structural design. Those things have contributed to increasing emergence of excessive vibration on structures, especially on slabs [2,3]. This problem is directly linked to the structural control devices study. Structural control provide changes on stiffness and damping properties of the system, installing external devices or applying external forces, aiming to reduce original structure vibrations.

Among several structural control devices there is the Tuned Mass Damper (TMD), a damper applied in many practical cases, such as high buildings, footbridges, towers, etc. [4]. It is a mass-spring-dashpot system connected to the structure in a way to vibrate out of phase with the main structure, if tuned to the appropriated frequency. In this way the energy is transferred to the TMD reducing amplitude vibration on the main system. The use of TMD to reduce floor vibration has been studied by many researchers lately [5-9]; however there are still many topics to further development, such as the determination of appropriate TMD parameters to optimize control performance. This work presents a floor vibration study to a practical case of a gym. The slab analyzed presented excessive vibration problems when subjected to typical human induced loading of rhythmic exercise practicing. A control system consisting of various TMD was proposed. This proposal was based on a parametrical study to improve the performance of a slab set with typical dimensions of structures like residential and commercial buildings.

2. HUMAN INDUCED VIBRATIONS

Nowadays the design of building floors with low natural frequencies, with values near of those from human induced loads, can cause undesirable excessive vibrations.

Thus the understanding of structural dynamic behavior and the characteristics of the dynamic load is very important, particularly to floor slabs. Dynamic loading can produce high vibration levels than can cause discomfort to users and even compromise structural safety. Occurrence's of these problems are well known worldwide, like the already mentioned London Millenium Footbridge [1]. The stadium grandstands collapse in the 1920 years lead the American Standards Association (ASA, atual ANSI—American National Standards Institute) to nominate a committee to establish safety standards for this type of structure. Decades later, researchers described mathematically human loading arising from typical activities using a Fourier series. Faisca [10] performed an experimental study focusing on this type of load, describing human induced dynamic load considering loss of contact with the structure. According to her, these loads can change depending on structure's flexibility, namely they depend on people-structure interaction. Varela [5] studied human induced loading correlating the mathematical model with experimental results.

3. STRUCTURAL CONTROL: TUNED MASS DAMPER (TMD)

Tuned Mass Damper reduces the energy dissipation request of structural members when subjected to dynamic loads. This reduction occurs by transferring a portion of the vibratory energy to the TMD that in its simplest conception is a mass-spring-dashpot system connected to the structure, like shown on **Figure 1**. Considering the main system like a single-degree-of-freedom model, with mass M, damping C and stiffness K, on which is applied a dynamic load f(t) and attached a TMD with mass m, damping c and stiffness k, the equations of motion are:

$$M\ddot{y}(t)+C\dot{y}(t)+Ky(t)=c\dot{z}(t)+kz(t)+f(t) \tag{1}$$

$$m\ddot{z}(t)+c\dot{z}(t)+kz(t)=-m\ddot{y}(t) \tag{2}$$

where:

$\ddot{y}(t)$: main system acceleration;

$y(t)$: main system velocity;

$y(t)$: main system displacement;

z(t): relative displacemente between the TMD and the main system.

When installed, the TMD tries to bring resonance peaks to lower values; this effect is desirable occurring on a wide frequency range.

For this to happen, TMD optimum parameters should be obtained, improving the control system performance [11]. A large number of these devices have been installed in high buildings, bridges, towers and industrial chimneys to control the dynamic response due to strong winds mostly [2]. In recent years studies were developed to harness the potential of TMD on floor vibration control. Battista & Varella [12] presented a practical case of using TMDs in commercial building with metallic beams and composite floor decks that presented excessive vibration levels according to ISO 2631/1, ISO 2631/2 and NBR 8800 ABNT codes. Setareh et al. [8] describe the application of a passive pendulum working as a TMD for two cases with excessive slab vibrations. In this study they emphasize men-structure interaction on experimental results obtained.

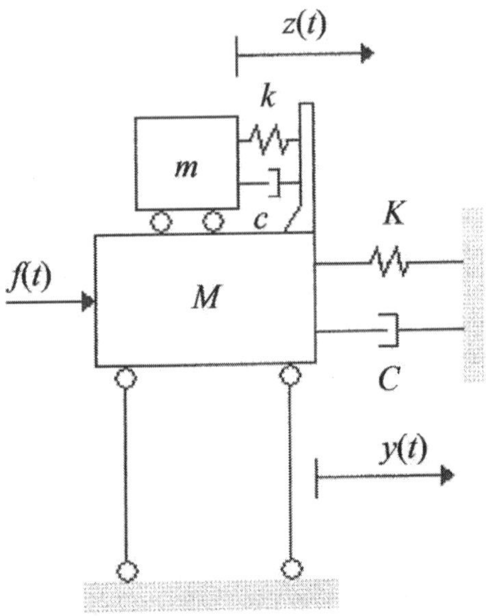

Figure 1. Structural system model with a TMD installed [4].

4. TMD PARAMETRIC STUDY

The four slab structures studied in this work are shown in **Figure 2**. The loading applied simulated people on the move.In all cases analyzed it was verified vibration levels higher than those recommended on standard codes. Thus it was performed a study analyzing different TMD configurations in a way to verify the best control proposal for each one of the slabs. This study also analyzed the influence of TMD parameters on the control strategy performance. Because of the absence of design criteria to set parameters to TMD installed on floors, it was performed a parametrical study varying the mass ratio μ, the frequency ratio α and the damping ratio ξ, and verifying for each set of parameters the TMD efficiency. The four models presented natural frequencies near of values that characterize human dynamic loads. On the first slab studied, simply supported at the four edges, the best configuration obtained was a single TMD placed at the midspan, it reduced on 75% displacement amplitude.

Table 1 presents the bests reductions for each slab, applying harmonic loading and human-induced loading, associated frequency ratios are also presented. It can be verified that best reductions were obtained considering the theoretical harmonic loading compared to human-induced loading. In the case of Slab 1, the best verified value of α was 0.95, namely tuning the TMD very close to the corresponding slab natural frequency. For damping ratio it was concluded that increasing its value can worsen TMD performance. The more increase applied to mass ratio, best is the efficiency. However, this parameter requires special attention because the damper weight increase can lead to excessive static deformations, compromising structural safety. The parametric studies performed allowed to observe the influence of the frequency ratio (α) change in the displacement reducing, shown in**Figure 3**, as well as mass ratio (μ) and damping ratio influence (ξ), represented on Figures 4 and 5. Similar studies were performed by in the other slabs shown in **Figure 2**, they are well detailed on the work of Santos [13]. Numerical simulations indicated that the parameter which design should be more careful, since it has a great influence on final results, is the frequency ratio α. Damping ratio ξ and mass ratio μ didn't show the same influence.

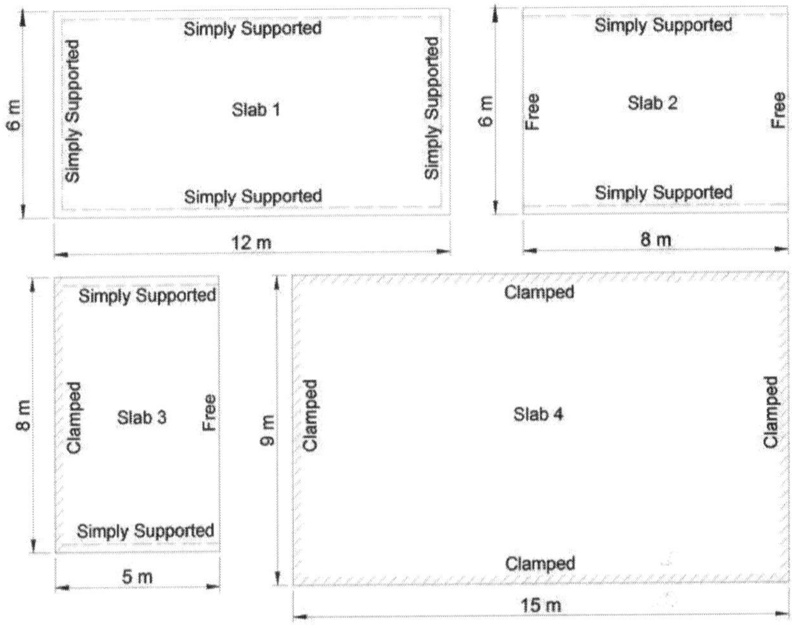

Figure 2. Dimensions and support conditions of the four slab models analyzed.

5. ANALYSIS OF A GYM SLAB

A gym floor, that presented excessive vibrations, located in a commercial building situated in the city of Brasilia, Brazil, was analyzed via Finite Element Method using ANSYS software. The commercial building has five floors: a basement, downstairs, first and second floor and roof. All floors were built as concrete solid slabs with 15 MP a resistance, 7 cm of thickness, supported by beams and columns of rectangular section.

Table 1. Parametrical study results: frequency ratio and displacement reduction.

	Type of loading	Number of TMDs	Position	α	Displ. reduction (%)
Slab 01	Harmonic	1	Midspan	0.95	74.58
	Human-induced 12 people	1	Midspan	0.98	45.70
Slab 02	Harmonic	2	1 m from the border	1.03	73.12
	Human-induced 8 people	2	At the border	1.10	54.65
Slab 03	Harmonic	1	0.8 m from the border	1.00	79.90
	Human-induced 8 people	1	At the border	1.25	67.6
Slab 04	Harmonic	1	Midspan	0.95	72.09
	Human-induced 8 people	1	Midspan	1.00	56.00

A rectangular portion of the floor situated in downstairs floor with 14.074 m × 6.49 m and 3.15 m of height, showed on **Figure 6** was considered here. In this sector of the floor is located the ballroom gymnastics. The physical properties attributed to the numerical model were: Modulus of Elasticity of 29 Gpa; density of 2500 kg/m^3 and Poisson ratio of 0.3. The studied slab was modeled with a finite element mesh in the XY plane, using SHELL63 elements. Beams were modeled using BEAM4, as well as columns.

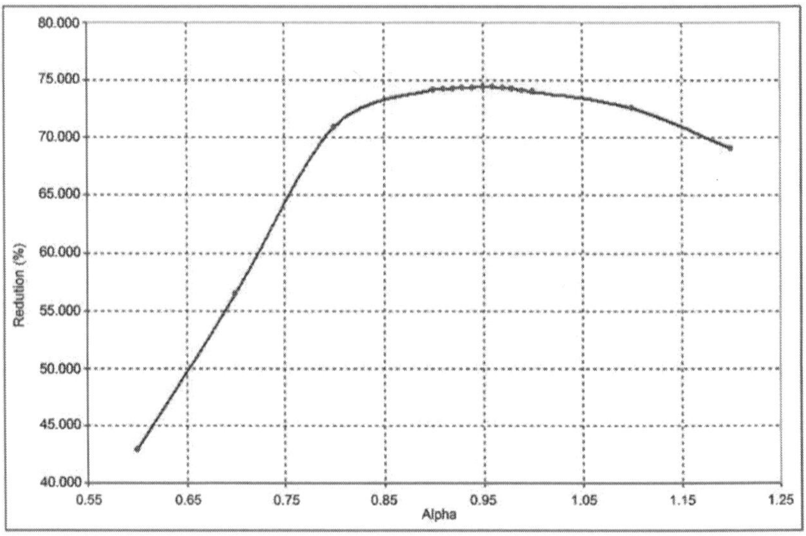

Figure 3. Frequency ratio α influence on displacement reduction (μ = 0.10 and ξ = 0.40) Slab 1.

Modal analysis showed natural frequencies values near of those from human induced loads. The first three mode shapes are in the horizontal direction with bending of the columns in the first mode and column bending and torsion in the second and third modes. From the fourth mode arise vertical displacements on the structure [13]. A transient analysis was also performed simulating the loading of people jumping. Nineteen people were considered placed at a distance of 1.0 m between one another, covering all the slab area.

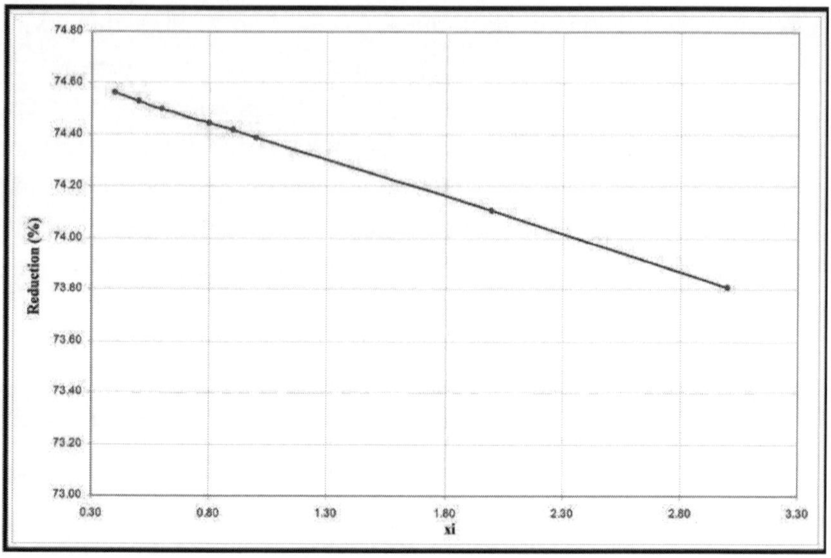

Figure 4. Damping ratio ξ influence on displacement reduction (μ = 0.10 and α = 0.96) Slab 1.

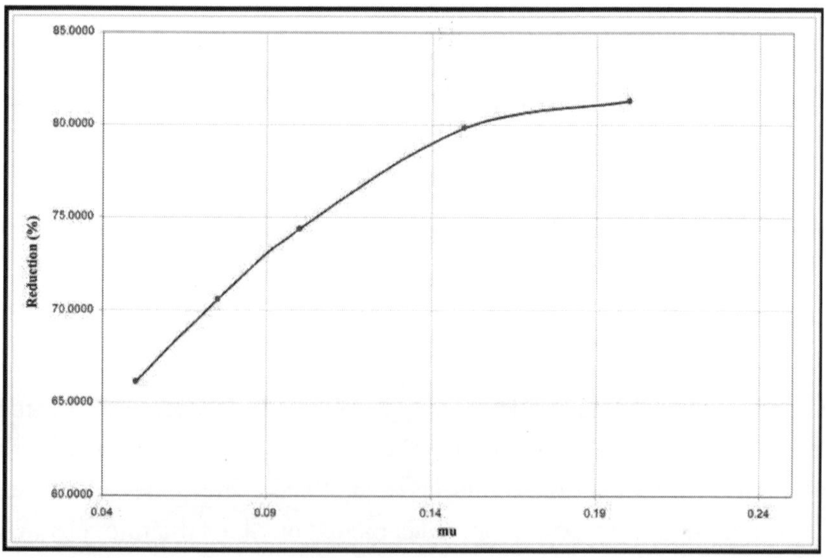

Figure 5. Mass ratio influence μ on displacement reduction (ξ = 0.4 and α = 0.96) Slab 1.

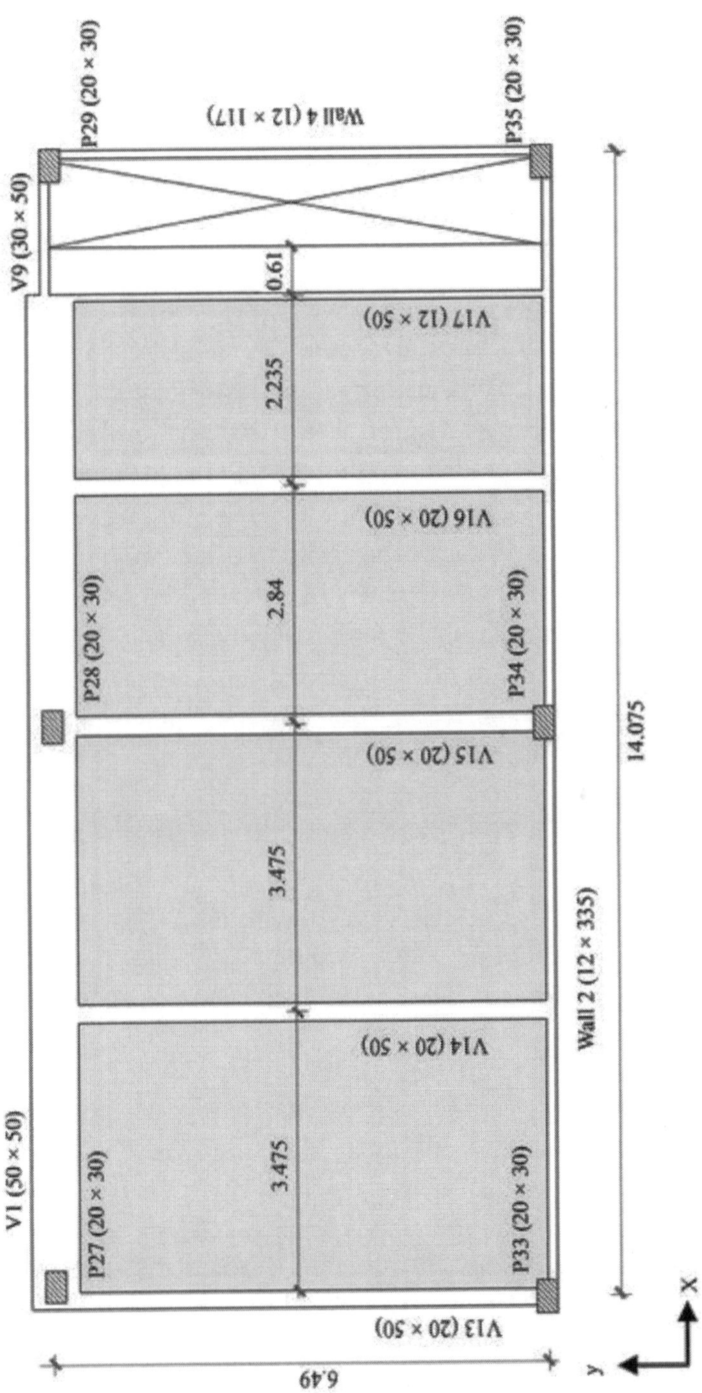

Figure 6. Dimensions and structural sketch of the gym floor portion studied.

Table 2 present the maximum response value on the node where it was verified the largest vertical displacement on the transient analysis of nineteen people jumping.

To attenuate these excessive vibrations it was proposed a control strategy using TMDs. In the absence of design TMD criteria, initially Jangid and Den Hartog [4,14] parameters were considered. It is noteworthy that these above parametric suggestions were based on different type of structures like single-degree-of-freedom model and shear buildings. The best performance was achieved using 6 TMD where three of them were tuned to the first natural frequency (2.56 Hz) and working on Y direction (horizontal) and the other ones tuned in the fourth natural frequency (9.35 Hz) and working on Z direction (vertical). The Z directions dampers were placed individually on the midspan of beams V14, V15 and V16. The Z direction dampers were located in the midspan of the three slabs; these nodes presented the largest displacements in the analysis. However, those were not good results, since it was achieved a 20% acceleration reduction, but also a 10% increase on displacements.

Therefore, a new analysis, based on the parametric studies presented earlier in this work, was performed.The nodes chosen to attach the TMDs were the slab's midspan nodes, which presented the largest displacement amplitudes. Three proposals with different numbers of TMDs were studied (1, 2 and 3 dampers). The best proposal showed to be the one using a single TMD.

The TMD was placed on the node in the slab midspan between beams V13 and V14, as shown in**Figure 7**. The adopted TMD parameters were a = 0.96; m = 0.10 and x = 0.4. The TMD was tuned to the first and fourth frequencies (A1 and A2 models) as shown on **Table 3**. Maximum responses are presented on **Table 4**. Model A1 was the better alternative with a 10.449% displacement reduction and a 12.995% acceleration reduction. Despite presenting a greater acceleration reduction, model A2 presented displacement amplification.

Table 2. Uncontrolled largest node response.

No.	Uz (m)	zV (s/m)	zA (s^2/m)
472	0.790	0.0172	0.6694

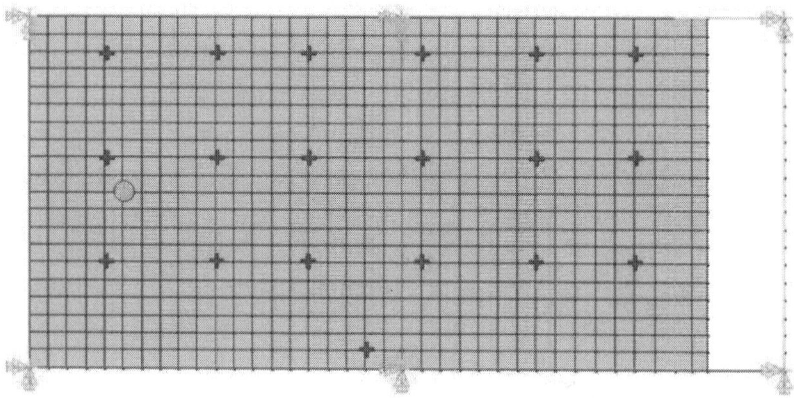

Figure 7. Gym slab with TMD installed on node 386, better position to vibration reduction.

Still considering a single TMD placed at the same node, a new set of parameters was attributed based on work of Santos [13] parametric studies of a 3D model considering beams and columns, similar to the present work structure. The parameter values adopted were $\alpha = 0.82$, m = 0.10 and x = 0.4, as indicated on **Table 5**, where A3 is the model tuning the TMD on the first frequency and A4 the TMD tuned on the fourth frequency. The dynamic response in both cases is presented onTable 6.

It is verified that, as well as in the last proposal, a better performance with the TMD tuned on the first frequency presenting a 7.514% displacement reduction and a 9.881% acceleration reduction. Nevertheless, model A1 showed better performance.

Table 3. TMD properties to models A1 and A2.

Model	No. AMS	f_{est} (Hz)	M_{est} (kg)	μ	ξ	α	ω_{est} (rad/s)	ω_{AMS} (rad/s)	k_{AMS} (N/m)	m_{AMS} (kg)	C_{AMS} (N·s/m)
A1	1	2.56	45034.7	0.10	0.004	0.96	16.08	15.44	1073751.52	45034.7	556.31
A2	1	9.35	45034.7	0.10	0.004	0.96	16.08	56.40	14323431.69	45034.7	2031.83

Table 4. Maximum displacements, accelerations and corresponding reductions to A1 and A2 models.

Model	Node		Undamped	Damped	Reduction Δ%
A1	386	U_z (mm)	0.712	0.638	10.364
		A_z (m/s²)	0.6214	0.541	12.955
A2	386	U_z (mm)	0.712	0.827	−16.15
		A_z (m/s²)	0.6214	0.494	20.56

Table 5. TMD properties to models A3 and A4.

Model	No. AMS	f_{est} (Hz)	M_{est} (kg)	μ	ξ	α	ω_{est} (rad/s)	ω_{AMS} (rad/s)	k_{AMS} (N/m)	m_{AMS} (kg)	C_{AMS} (N·s/m)
A3	1	2.56	45034.7	0.10	0.004	0.82	16.08	13.19	783409.85	4503.47	475.18
A4	1	9.35	45034.7	0.10	0.004	0.82	48.17	48.17	10450385.71	4503.47	1735.52

Table 6. Maximum displacements, accelerations and corresponding reductions to A3 and A4 models.

Model		Node	Undamped	Damped	Reduction Δ%
A3	386	U_z (mm)	0.712	0.6585	7.514
		A_z (m/s²)	0.6214	0.6500	9.881
A4	386	U_z (mm)	0.712	0.7817	−9.789
		A_z (m/s²)	0.6214	0.4376	29.578

Table 7. Comparative results of displacements and accelerations.

Proposals	No. AMS	Uz (mm) Δ%	Az (m/s^2) Δ%
A1 model	01	10.449	12.955
6 TMD proposal	06	−8.421	23.887

Based on all presented results it can be concluded that the displacement reduction is better achieved tuning the TMD to the first frequency. In acceleration reduction the best performance happened when tuning the TMD to the fourth frequency, but in these cases an undesirable increase on displacements was observed. In this way, the best proposal to control the structure is model A1, using a single TMD tuned to the first natural frequency of the structure. A comparison between model A1 and the 6 TMD initial proposal is shown in **Table 7**. Model A1 was the more indicated design in a way that despite not reducing so well accelerations, no increase in displacement is noticed. A displacement increase is not recommendable in a structural point of view.

6. CONCLUSION

This paper presents the study of the vibration problem of a commercial building floor used as a gym. The concrete slab analyzed presented excessive vibrations when subjected to typical exercise rhythmic loading. A control system was proposed using the Tuned Mass Damper device. Also a parametric study to design the TMD was performed considering four different concrete slabs with varying dimensions, support conditions and loads. The control system showed to be effective reducing by 10% displacement amplitudes of the gym floor analyzed.

ACKNOWLEDGEMENTS

This work has been supported by CNPq and CAPES brazilian agencies which are gratefully acknowledged.

REFERENCES

1. P. Dallard, A. I. Fitzpatrick, A. Flint, S. Le Bourva, A. Low, R. Smith and M. R. Willford, "The London Millennium Footbridge," The Structural Engineer, Vol. 79, No. 22, 2001, pp. 17-33.

2. I. Saidi, "Development of a Viscoelastic Tuned Mass Damper to Reduce Walking Induced Vibrations in Building Floors," Thesis of Doctor of Philosophy, Swinburne University of Technology, Melbourne, 2012.

3. J. E. Campuzano, R. Castro, S. M. Avila and G. Doz, "Platform for Dynamic Tests: Preliminary Studies, Design and Construction," RevistaIbracon de Estruturas e Materiais, Vol. 6, No. 1, 2013, pp. 55-74.

4. T. T. Soong, G. F. Dargush, "Passive Energy Dissipation in Structural Engineering," John Wiley & Sons, Chichester, 1997.

5. W. D. Varela, "Modelo Teórico Experimental para Aná- lise de Vibrações Induzidas por Pessoas Caminhando Sobre Lajes de Edifícios," Ph.D. Thesis, COPPE/UFRJ, Rio de Janeiro, 2004.

6. D. V. F. Lima, "Controle de Vibrações Induzidas em uma Laje de Academia de Ginástica com a Utilização de Amortecedores de Massa Sintonizados," Master Thesis, Universidade de Brasília, Brasília, 2007.

7. M. Setareh, J. K. Ritchey, A. J. Baxter and T. M. Murray, "Pendulum Tuned Mass Dampers for Floor Vibration Controll," Journal of Performance of Constructed Facilities, ASCE, Vol. 20, No. 1, 2006, pp. 64-73.

8. M. Setareh, J. K. Ritchey, T. M. Murray, J. H. Koo and M. Ahmadian, "Semiactive Tuned Mass Dampers for Floor Vibration Control," Journal of Structural Engineering, ASCE, Vol. 133, 2007, pp. 242-250.

9. W. D. Varela and R. C. Battista, "Control of Vibration Induced by People Walking on Large Span Composite Floor Decks," Engineering Structures, Vol. 33, No. 9, 2011, pp. 2485-2494.

10. R. G. Faisca, "Caracterização de Cargas Dinâmicas Geradas por Atividades Humanas," Ph.D. Thesis, COPPE/ UFRJ, Rio de Janeiro, 2003.

11. S. M. Avila, "Controle Híbrido para Atenuação de Vibra- ções em Edifícios," Ph.D. Thesis, Pontifícia Universidade Católica do Rio de Janeiro, Rio de Janeiro, 2002.

12. R. C. Battista and W. D. Varela, "Medidas Corretivas para Vibrações de Painéis Contínuos de Lajes de Edifícios," Proceedings of the 30th Jornadas Sul-Americanas de Engenharia Estrutural, Brasilia, 2002.

13. M. D. S. Santos, "Análise Numérica Do Controle de Vibrações em Lajes de Edifícios Utilizando Amortecedores de Massa Sintonizados," Master Thesis, Universidade de Brasília, Brasília, 2009.

14. R. Jangid, "Optimum Multiple Tuned Mass Dampers for Base Excited Undamped System," Earthquake Engineering and Structural Dynamics, Vol. 28, No. 9, 1999, pp. 1041-1049.

CHAPTER 3

Vibration Analysis of Composite Laminate Plate Excited by Piezoelectric Actuators

Shiuh-Chuan Her * and Chi-Sheng Lin

Department of Mechanical Engineering, Yuan Ze University, Chung-Li 320, Taiwan

ABSTRACT

Piezoelectric materials can be used as actuators for the active vibration control of smart structural systems. In this work, piezoelectric patches are surface bonded to a composite laminate plate and used as vibration actuators. A static analysis based on the piezoelectricity and elasticity is conducted to evaluate the loads induced by the piezoelectric actuators to the host structure. The loads are then employed to develop the vibration response of a simply supported laminate rectangular plate excited by piezoelectric patches subjected to time harmonic voltages. An analytical solution of the vibration response of a simply supported laminate rectangular plate under time harmonic electrical loading is obtained and compared with finite element results to validate the present approach. The effects of location and exciting frequency of piezoelectric actuators on the vibration response of the laminate plate are investigated through a parametric study. Numerical results show that modes can be selectively excited, leading to structural vibration control.

KEYWORDS

piezoelectric actuator; composite laminate; smart structure; harmonic vibration

1. INTRODUCTION

Smart structures incorporating active devices to sense and actuate the structure could be applied in many engineering applications, such as aircraft structures, satellites, large space structures and so forth [1,2]. Many materials have been tested as actuators and sensors such as piezoelectric materials, shape memory alloys, electrostrictive materials, magnetostrictive materials, electro-rheological fluids, and fiber optics [3]. These materials can be embedded into or surface bonded with structures, thus acting as either a sensor or an actuator. Among them, piezoelectric materials have become popular because of high strength, temperature insensitivity, and ease of implementation. Active vibration control of flexible structures using piezoelectric sensors and actuators has received much attention in recent years. Successful implementation of active adaptive control is usually based on the full understanding of the smart structure system, which significantly affects the control strategy, parameter tunings and convergence [4]. This process has proved to be tedious and difficult if it is done only by experiments. Thus, suitable simulation models are necessary to facilitate the development. Numerous efforts have been made in the last few years to develop suitable simulation models. Alibeigloo and Madoliat [5] presented a three-dimensional solution for static analysis of a cross-ply rectangular plate imbedded in piezoelectric layers using the differential quadrature method (DQM) and a Fourier series approach. Proulx and Cheng [6] reported a dynamic model for the active vibration control of a rectangular plate using piezoceramic elements of arbitrary shape. Susanto [7] presented an analytical model of piezoelectric laminated slightly curved beams, which included the computation of natural frequencies, mode shapes and transfer function formulation using the distributed transfer function method. Qiu *et al.* [8] used piezoelectric ceramic patches as sensors and actuators to suppress the vibration of a smart flexible clamped plate. An optimal placement method for the locations of piezoelectric actuators and sensors was developed based on the degree of observability and controllability indices for the cantilever plate. Santos *et al.* [9] developed a finite element model for the analysis of 3D axisymmetric laminated shells with piezoelectric sensors and actuators. Fernandes and Pouget [10] predicted the static and dynamic (vibration) responses of composite plates excited by piezoelectric actuators based on the kinematic assumption of the Love-Kirchhoff thin plate theory. The influence of the actuator position on the global and local responses of the composite plate was illustrated. Qing *et al.* [11] employed a modified mixed variational principle for piezoelectric materials to establish a semi-analytical solution for static and dynamic analysis of plates with piezoelectric patches. Huang and Sun [12] studied the load transfer and wave propagation of an anisotropic elastic medium induced by the surface bonded piezoelectric actuator. Dimitriadis *et al.* [13] used two dimensional patches of piezoelectric material bonded to the surface of a simply supported plate as vibration actuators to excite the selected modes. Qu *et al.* [14] proposed a dynamical model using the principle of minimum energy to study the vibration behavior of a piezoelectric composite plate with cracks, the effects of cracks and piezoelectric materials on mode shapes was presented. Della and Shu [15] presented a mathematical model basing on the Euler-Bernoulli beam theory and Rayleigh-Ritz approximation technique for the vibration of beams with embedded arrays of piezoelectric sensors and actuators. Smart structures often involve a coupling model between the host structure and piezoelectric sensors and actuators. In fact, a piezoelectric actuator introduces material and geometric discontinuities which lead to some mathematical difficulties. Because of the difficulties associated with the complicated electromechanical coupling, material inhomogeneity and anisotropy, analytical solutions representing the vibration behavior of a composite plate excited by PZT actuators have not been properly established. Most of the existing works focused on the numerical solutions such as finite element method. In this study, a theoretical model accomplished with analytical solution is presented to show how patch type piezoelectric actuators can be used to excite and control the vibration modes of a composite laminate plate.

The present work investigated the vibration response of a simply supported composite laminate plate excited by piezoelectric actuators. The model consists of two piezoelectric patches symmetrically surface bonded on a cross-ply composite laminate. Static analysis leads to the determination of the loads induced by the piezoelectric actuators to the host structure. This is followed by a dynamic analysis for a composite laminate plate excited by piezoelectric actuators with time harmonic electrical loading. A closed form solution to the harmonic vibration of the simply supported composite laminate was obtained by using the classical laminate plate theory. The analytical solution was validated by the corresponding finite element results. Results are presented for the vibration displacement distribution and modal shape of the composite laminate plate. The effects of excitation frequency and actuator location on the vibration displacement and modal shape are investigated through a parametric study.

2. BENDING MOMENT

In this work, two piezoelectric actuators are symmetrically bonded on the top and bottom surfaces of a cross-ply composite laminate. The polarized direction is along the z-axis. For an unconstrained thin piezoelectric actuator, equal strains in both x and y directions will be induced when activated by a voltage along the poling direction. The magnitude of the strain can be expressed in terms of the piezoelectric constant d_{31}, applied voltage V and actuator thickness t_{pe}, as follows:

$$(\varepsilon_x)_{pe} = (\varepsilon_y)_{pe} = \varepsilon_{pe} = \frac{d_{31}}{t_{pe}} V \tag{1}$$

where subscripts pe and p represent the quantities associated with piezoelectric actuator and host plate, respectively, throughout this paper. When an electrical field is applied in the direction normal to the actuator surface, surface strains are generated Equation (1). Due to the coupling of the actuator to the structure, forces and moments are induced in the bonded area of the structure. Since this work focuses on the deformation of the plate induced by the bending moment, only constantd_{31} is considered in this model.

The two actuators are activated by applying a voltage of equal magnitude and opposite sign to the opposing actuators. The opposite directions of the surface tractions at the interfaces between the actuator and plate cause the uniform bending moments along the actuator boundaries as shown in Figure 1.

The bending moments per unit length m_x and m_y induced by the actuators to the cross-ply composite laminate have been derived in the previous work [16] as follows:

$$m_x = C_1 \varepsilon_{pe} \tag{2a}$$

$$m_y = C_2 \varepsilon_{pe} \tag{2b}$$

$$C_1 = \left(A_1 (D_{11})_p + A_2 (D_{12})_p\right) \tag{3a}$$

$$C_2 = \left(A_1 (D_{12})_p + A_2 (D_{22})_p\right) \tag{3b}$$

$$A_1 = \frac{B_{11pe}(1+\nu_{pe})}{(D_{11})_p + D_{11pe}} + \frac{\left(\left((D_{12})_p + D_{12pe}\right)\left(-B_{11pe}\left((D_{11})_p + D_{11pe}\right)\left(1+\nu_{pe}\right) + B_{11pe}\left((D_{12})_p + D_{12pe}\right)\left(1+\nu_{pe}\right)\right)\right)}{\left(\left((D_{11})_p + D_{11pe}\right)\left(-\left((D_{12})_p + D_{12pe}\right)^2 + \left((D_{11})_p + D_{11pe}\right)\left((D_{22})_p + D_{22pe}\right)\right)\right)} \tag{4a}$$

$$A_2 = \frac{-B_{11pe}\left((D_{11})_p + D_{11pe}\right)\left(1+\nu_{pe}\right) + B_{11pe}\left((D_{12})_p + D_{12pe}\right)\left(1+\nu_{pe}\right)}{-\left((D_{12})_p + D_{12pe}\right)^2 + \left((D_{11})_p + D_{11pe}\right)\left((D_{22})_p + D_{22pe}\right)} \tag{4b}$$

Where B_{11}, D_1, D_{12}, D_{22} are related to the stiffness of the piezoelectric material and composite material.

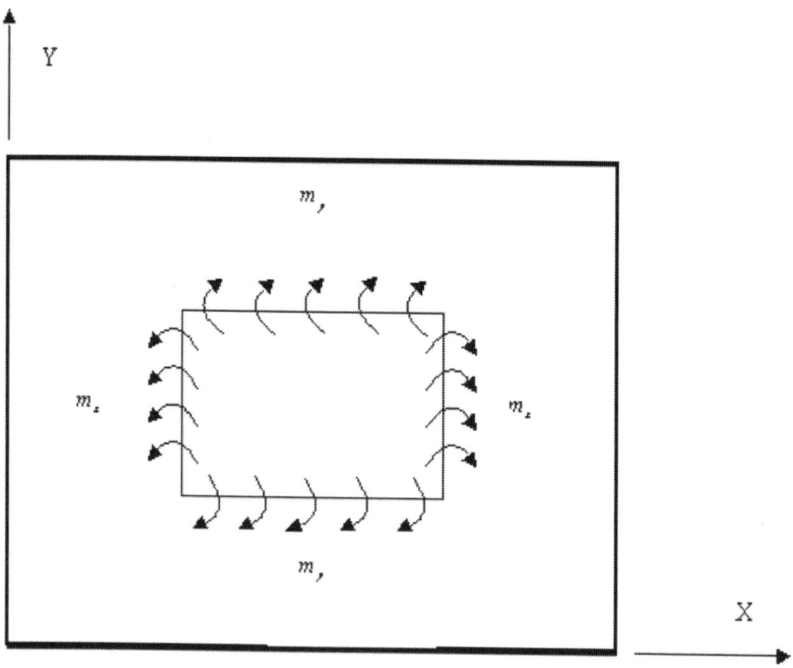

Figure 1. Bending moment acting on the composite laminate induced by piezoelectric actuators.

3. HARMONIC VIBRATION OF A SIMPLY SUPPORTED COMPOSITE PLATE

The host plate is a rectangular plate with simply supported boundary conditions. The location of the surface bonded actuators viewed from the top is shown in Figure 2. The activated piezoelectric actuators will induce bending moments as shown in Equation (2) to the composite plate and can be expressed in terms of unit step functions as follow:

$$m_x = C_1\varepsilon_{pe}\left[h(x-x_1)-h(x-x_2)\right]\left[h(y-y_1)-h(y-y_2)\right] \quad (5a)$$

$$m_y = C_2\varepsilon_{pe}\left[h(x-x_1)-h(x-x_2)\right]\left[h(y-y_1)-h(y-y_2)\right] \quad (5b)$$

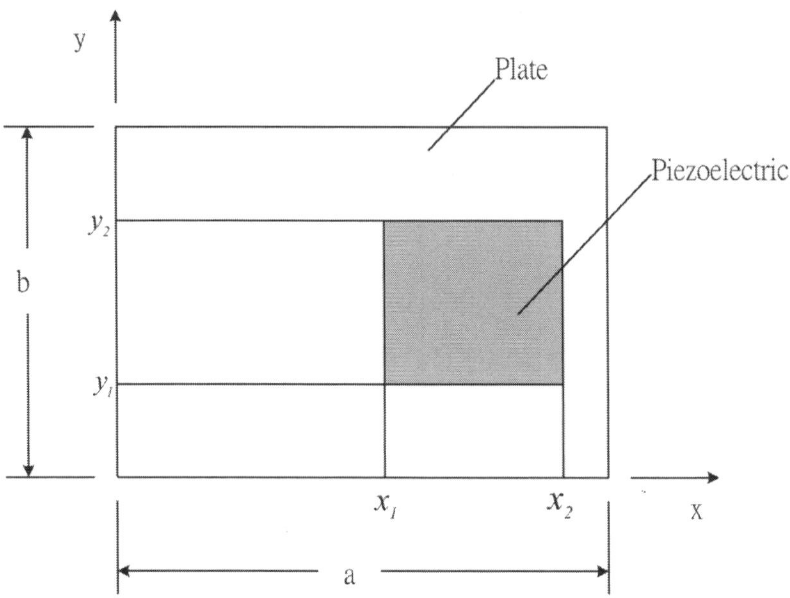

Figure 2. Surface bonded actuator on the composite laminate plate.

If the actuator input voltage is sinusoidal oscillating with frequency *p*, the bending moments m_x and m_y will oscillate at the same frequency as the voltage. Using the classical laminate plate theory, the equation of motion for the plate can be written in terms of the displacement w and the actuators induced moments m_x and m_y as:

$$(D_{11})_p \frac{\partial^4 w}{\partial x^4} + 2H_1 \frac{\partial^4 w}{\partial x^2 \partial y^2} + (D_{22})_p \frac{\partial^4 w}{\partial y^4} + m'' \ddot{w} = F(x,y) \sin pt \tag{6}$$

$$H_1 = (D_{12})_p + 2(D_{66})_p$$

$$F(x,y) = \frac{\partial^2 m_x}{\partial x^2} + \frac{\partial^2 m_y}{\partial y^2} = C_1(\varepsilon_{pe})[\delta'(x-x_1) - \delta'(x-x_2)][h(y-y_1) - h(y-y_2)]$$
$$+ C_2(\varepsilon_{pe})[h(x-x_1) - h(x-x_2)][\delta'(y-y_1) - \delta'(y-y_2)]$$

Where $(D_{11})_p$, $(D_{22})_p$, $(D_{66})_p$ are the bending stiffness of the composite laminate, m'' is the area mass density of the composite plate.

The forcing function $F(x, y)$ can be expressed in terms of a Fourier series as follows:

$$F(x,y) = \left(\sum_{m=1}^{\infty} \sum_{n=1}^{\infty} F_{mn} \sin \frac{m\pi x}{a} \sin \frac{n\pi y}{b} \right)$$

$$F_{mn} = \frac{4}{a*b} \int_0^b \int_0^a F(x,y) \sin \frac{m\pi x}{a} \sin \frac{n\pi y}{b} dxdy$$
$$= \frac{4}{a*b} \left[-\frac{C_2 \varepsilon_{pe} \gamma_m^2 + C_1 \varepsilon_{pe} \gamma_n^2}{\gamma_m \gamma_n} (\cos \gamma_m x_1 - \cos \gamma_m x_2)(\cos \gamma_n y_1 - \cos \gamma_n y_2) \right]$$

$$\gamma_m = \frac{m\pi}{a} \; ; \; \gamma_n = \frac{n\pi}{b} \tag{7}$$

For a simply supported rectangular plate, the flexural displacement w can be expressed by the following Fourier series:

$$w(x,y,t) = \sum_{m=1}^{\infty} \sum_{n=1}^{\infty} W_{mn} \sin \frac{m\pi x}{a} \sin \frac{n\pi y}{b} \sin pt \tag{8}$$

Substituting Equations (7) and (8) into Equation (6), yields:

$$W_{mn}\left(D_{11}\frac{m^4\pi^4}{a^4}+2H_1\frac{m^2\pi^2}{a^2}\frac{n^2\pi^2}{b^2}+D_{22}\frac{n^4\pi^4}{b^4}-m''p^2\right)=F_{mn} \quad (9)$$

The natural frequency of a simply supported composite plate is:

$$\omega_{mn}=\pi^2\sqrt{\frac{D_{11}\frac{m^4}{a^4}+2H_1\frac{m^2n^2}{a^2b^2}+D_{22}\frac{n^4}{b^4}}{m''}} \quad (10)$$

Substituting Equation (10) into Equation (9), results:

$$W_{mn}=\frac{F_{mn}}{m''\left(\omega_{mn}^2-p^2\right)} \quad (11)$$

Thus, the vibration displacement of the composite plate can be obtained by substituting Equation (11) into Equation (8) as follows:

$$w(x,y,t)=\left[\sum_{m=1}^{\infty}\sum_{n=1}^{\infty}\frac{F_{mn}}{m''\left(\omega_{mn}^2-p^2\right)}\sin\frac{m\pi x}{a}\sin\frac{n\pi y}{b}\right]\sin pt \quad (12)$$

4. FINITE ELEMENT ANALYSIS

The finite element method is a widely used and powerful tool for analyzing complex structures. Many researchers have modelled the piezoelectric actuation using the finite element method. The commercially available finite element software ANSYS has the ability to analyze piezoelectric materials. In this study, ANSYS is adopted to investigate the harmonic vibration of a simply supported composite plate excited by the surface bonded piezoelectric actuators. To perform the ANSYS finite element analysis for the piezoelectric actuator bonded structure, SOLID 45 and SOLID 5 elements were used in the composite plate and piezoelectric actuators, respectively.

A typical three dimensional finite element mesh is shown in Figure 3. A time harmonic voltage between the upper and lower surfaces of the SOLID 5 elements is applied, which results in an electric field along the poling direction of the actuator. The vibration displacements obtained from the finite element method are compared with the analytical solutions of Equation (12) to validate the present approach.

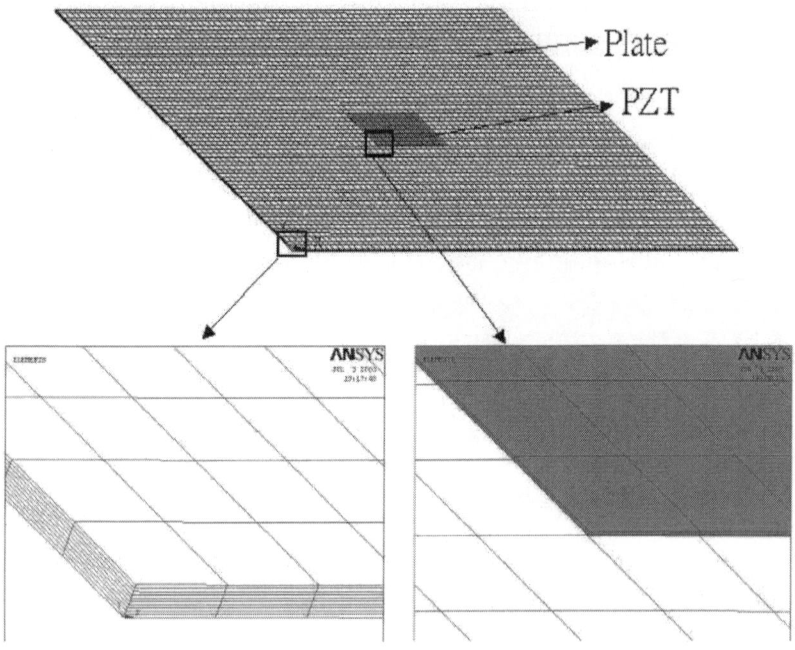

Figure 3. 3-D finite element mesh.

5. PARAMETRIC STUDY AND VERIFICATION

In the following numerical examples, the composite material is carbon/epoxy with stacking sequence [0/90/90/0].The composite material properties of the carbon/epoxy are listed in Table 1. The dimensions of the composite laminate plate are length a = 0.38 m, width b = 0.3 m, thickness t_p = 1.5876 mm. The piezoelectric actuator is assumed to be a PZT G-1195 with the following material properties [17]: Young's modulus E_{pe} = 63 GPa, Poisson's ratio v_{pe} = 0.3, density ρ_{pe} = 7,600 kg·m^{-2}, piezoelectric constant $d_{31} = 1.9 \times 10^{-10}$ V·m^{-1} and thickness t_{pe} = 0.15876 mm. The effects of the excitation frequency and location of the actuators are presented through a parametric study to investigate the vibration shape of the composite plate activated by the surface bonded piezoelectric actuators.

The natural frequencies of the simply supported composite laminate plate obtained by the finite element method and theoretical prediction Equation (10) are listed in Tables 2 and 3, respectively. It shows that the theoretical predictions agree well with the finite element results. The mode shapes of modes (1,1), (1,2), (2,1) and (2,2) are shown in Figures 4, 5, 6, and 7, respectively.

Example results are presented for the harmonic vibration of a rectangular composite thin plate excited by a pair of rectangular piezoelectric actuators with various dimensions and locations. To excite the selective vibration mode, the excitation frequency can be determined using Equation (10). Four different excitation frequencies (350, 650, 870 and 1,050 rads/s) were tested. Among these excitation frequencies, 350, 870 and 1,050 rads/s are close to the resonant frequencies of the (1,1), (1,2) and (2,1) modes, respectively, while the 650 rads/s is far away from those resonant frequencies. The vibration displacements calculated by the theoretical prediction (Equation (12)) are compared with the finite element results. In this work, the forcing function F(x, y,) as shown in Equation (7) was expressed in terms of Fourier series. Convergence test results show that 10 terms of Fourier series can converge to a correct solution. Thus, 10 terms of Fourier series was adopted throughout this analysis.

Table 1. Material properties of carbon/epoxy.

Longitudinal Modulus E_1	Transverse Modulus E_2	Shear Modulus G_{12}	Shear Modulus G_{23}	Poisson's Ratio ν_{12}	Poisson's Ratio ν_{23}
108 GPa	10.3 GPa	7.13 GPa	4.02 GPa	0.28	0.28

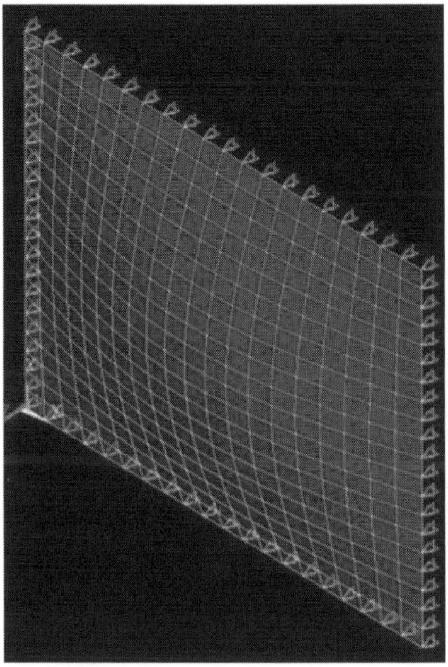

Figure 4. Mode shape of (1,1).

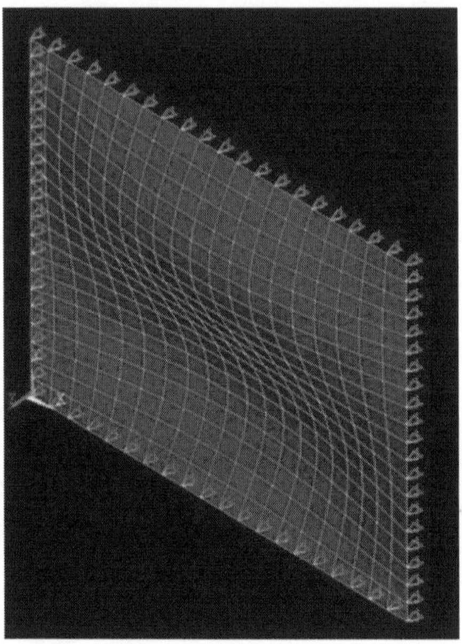

Figure 5. Mode shape of (1,2).

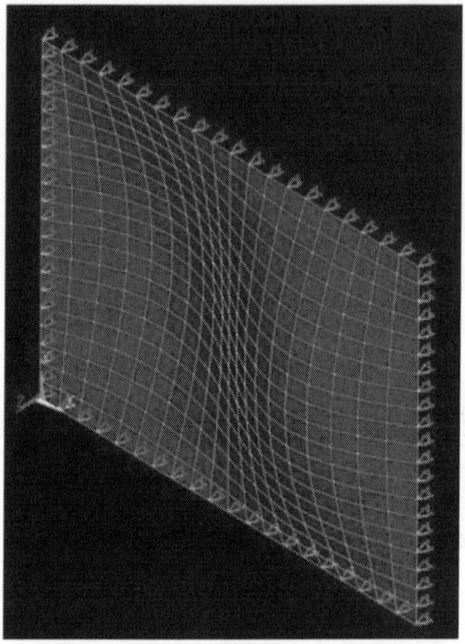

Figure 6. Mode shape of (2,1).

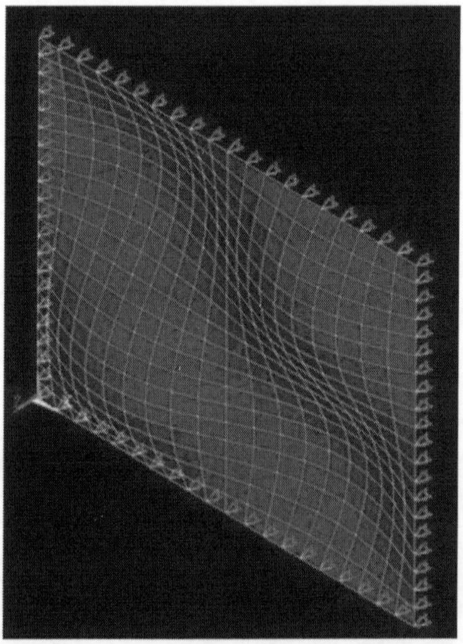

Figure 7. Mode shape of (2,2).

Table 2. Natural frequencies (rad/s) of the composite plate obtained by finite element method.

		n		
		1	2	3
m	1	358.5	875.3	1805.3
	2	1056.4	1433.6	2247.3
	3	2262.8	2561.8	3223.7

Table 3. Natural frequencies (rad/s) of the composite plate obtained by theoretical prediction Equation (10).

		n		
		1	2	3
m	1	359.1	876.3	1806.9
	2	1057.6	1436.2	2251.4
	3	2266.1	2566.8	3231.6

5.1. Three Different Sizes of Piezoelectric Actuators

Two piezoelectric actuators are surface bonded on the top and bottom surfaces of the composite plate. Three different sizes of piezoelectric actuators with the dimensions of 0.06 m × 0.04 m, 0.08 m × 0.06 m and 0.1 m × 0.08 m, respectively, bonded on the central area of the composite plate as shown in Figure 8 are considered in this example.

The piezoelectric actuators are excited by time harmonic voltages with three different frequencies of 350, 650 and 870 rads/s, respectively.

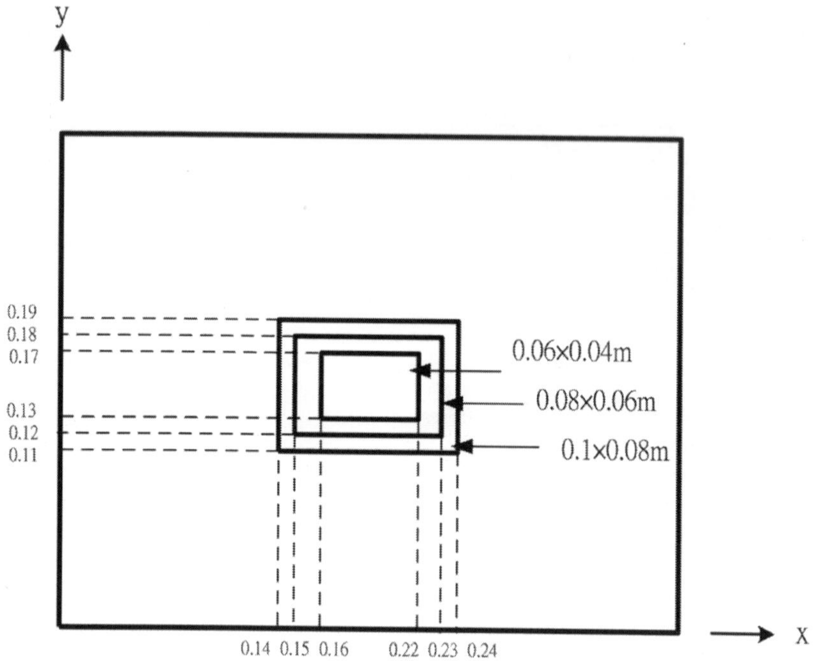

Figure 8. Three different sizes of PZT actuators.

5.1.1. Excitation Frequency *p = 350 rads/s*

Figure 9 shows the vibration profile of the composite plate excited by the PZT actuators with excitation frequency of 350 rads/s. As can be seen from Tables 2 and 3, this frequency of 350 rads/s is close to the resonant frequency of mode (1,1). It appears from Figure 9 that this mode is strongly excited with the vibration profile close to (1,1) mode shape. The displacement amplitudes along the horizontal line $y = b/2$ of the composite plate are presented in Figure 10. The vibration amplitude is increasing as the size of actuators increases. The vibration displacements of the composite plate predicted byEquation (12) and finite element method are in a good agreement. The maximum vibration amplitudes of the composite plate excited by the three different sizes of actuators are listed in Table 4. It shows that the difference between the present approach and finite element method is within 9%.

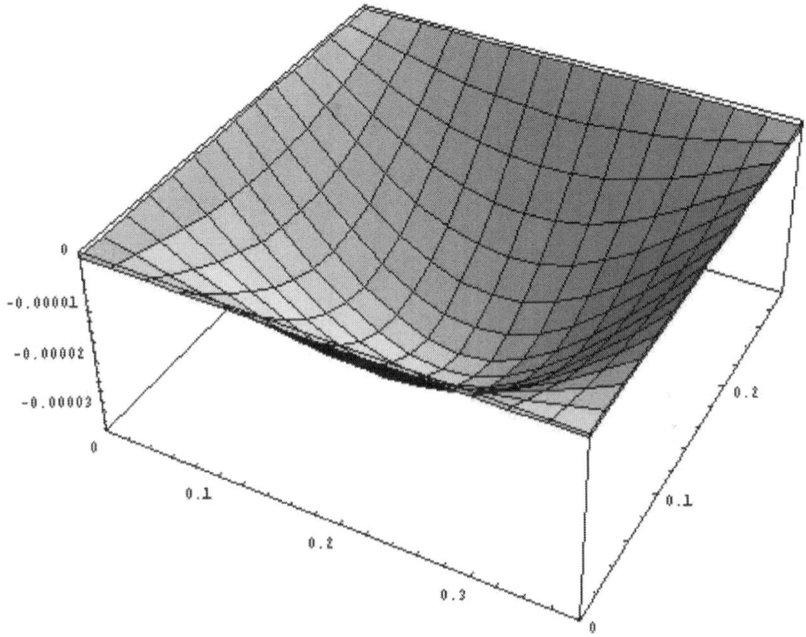

Figure 9. Vibration profile of the composite plate excited by PZT actuator bonded on the center of the plate with excitation frequency 350 rads/s.

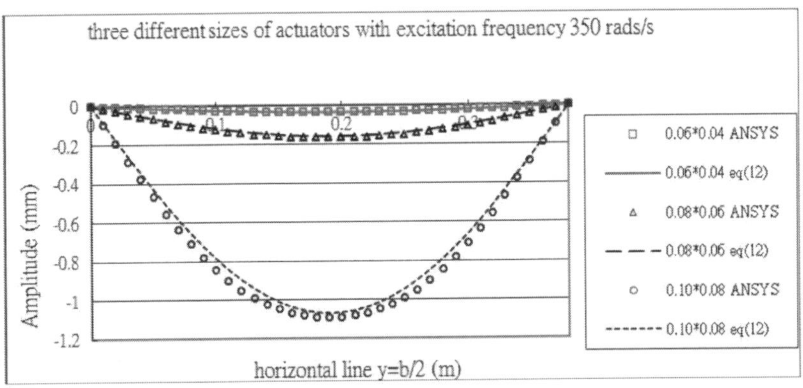

Figure 10. Vibration displacements of the composite plate obtained by ANSYS andEquation (12) along the horizontal line ($y = b/2$) excited by PZT actuator bonded on the center of the plate with excitation frequency 350 rads/s for three different sizes of actuators.

Table 4. Maximum vibration amplitudes (mm) excited by three different sizes of PZT actuator bonded on the center of the plate with excitation frequency 350 rads/s.

PZT Size	Finite Element ANSYS	Theoretical Prediction Equation (12)	Difference %
0.06 m × 0.04 m	3.53×10^{-2}	3.33×10^{-2}	6.0
0.08 m × 0.06 m	1.66×10^{-1}	1.56×10^{-1}	6.4
0.10 m × 0.08 m	1.15	1.06	8.5

5.1.2. Excitation Frequency *p = 870 rads/s*

Figure 11 plots the vibration profile of the composite plate excited by the PZT patches with excitation frequency of 870 rads/s. The frequency 870 rads/s is close to the resonant frequency of mode (1,2). Similar results can be observed that this mode is strongly excited with the vibration profile close to (1,2) mode shape. The displacement amplitude distribution along the horizontal line $y = b/2$ in Figure 12 illustrates the evidence of the mode (1,2) being excited.

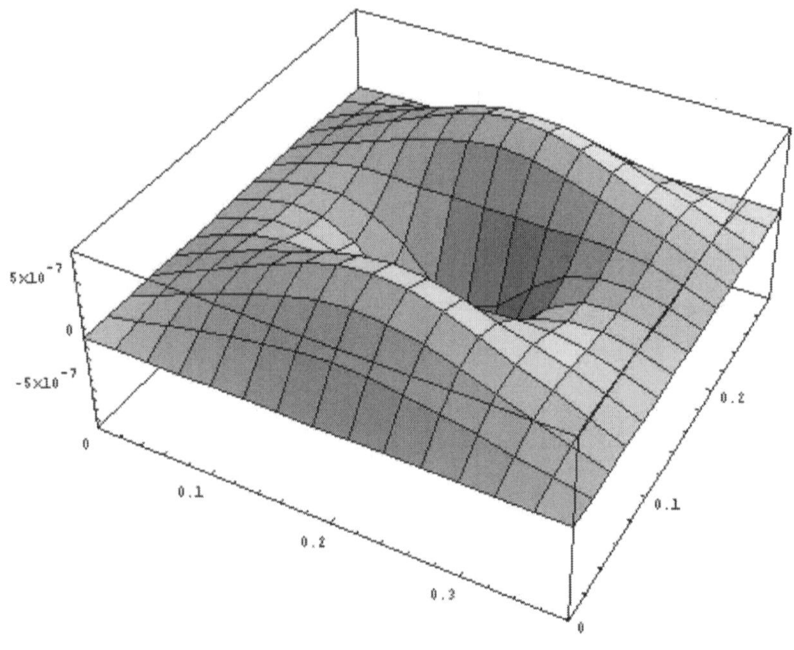

Figure 11. Vibration profile of the composite plate excited by PZT actuator bonded on the center of the plate with excitation frequency 870 rads/s.

The vibration amplitude is increasing as the size of actuators increases. The vibration displacements of the composite plate predicted by Equation (12) and finite element method are in a good agreement. The maximum vibration amplitudes of the composite plate excited by the three different sizes of actuators are listed in Table 5. It shows that the difference between the present approach and finite element method is within 9%.

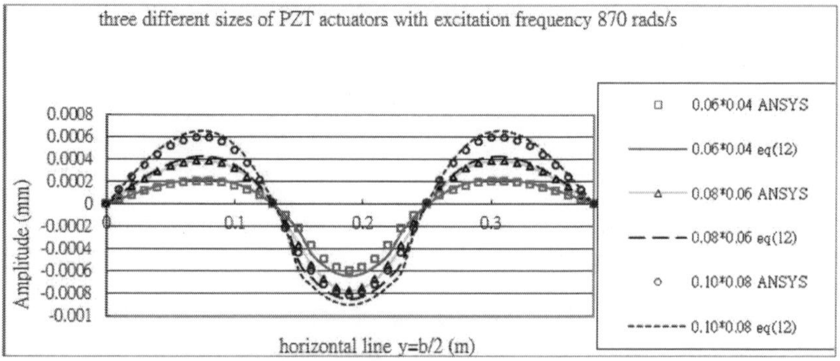

Figure 12. Vibration displacements of the composite plate obtained by ANSYS andEquation (12) along the horizontal line ($y = b/2$) excited by PZT actuator bonded on the center of the plate with excitation frequency 870 rads/s for three different sizes of actuators.

5.1.3. Excitation Frequency *p = 650 rads/s*

The excitation frequency was selected between modes (1,1) and (1,2), the system is being excited off the resonance. The displacement amplitude distribution along the horizontal line $y = b/2$ is presented in Figure 13. The vibration displacements of the composite plate predicted by Equation (12) and finite element method are in a good agreement. The maximum amplitudes of the composite plate excited by the three different sizes of actuators are listed in Table 6. It shows that the difference between the present approach and finite element method is within 5%.

Table 5. Maximum vibration amplitudes (mm) excited by three different sizes of PZT actuator bonded on the center of the plate with excitation frequency 870 rads/s.

PZT Size	Finite Element ANSYS	Theoretical Prediction Equation (12)	Difference %
0.06 m × 0.04 m	5.95×10^{-4}	6.42×10^{-4}	7.4
0.08 m × 0.06 m	7.74×10^{-4}	8.49×10^{-4}	8.8
0.10 m × 0.08 m	1.17×10^{-3}	1.12×10^{-3}	4.3

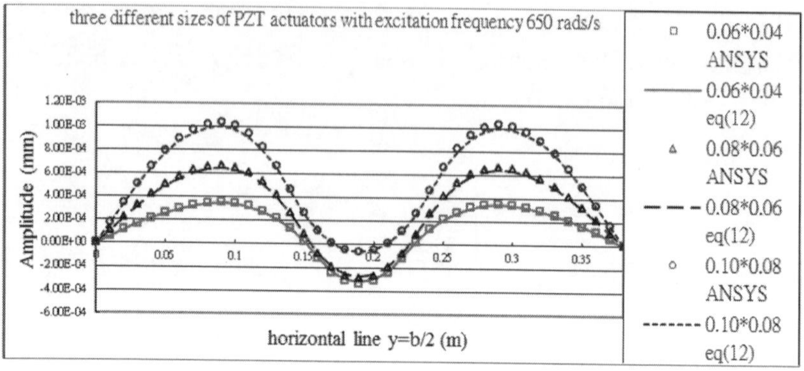

Figure 13. Vibration displacements of the composite plate obtained by ANSYS andEquation (12) along the horizontal line ($y = b/2$) excited by PZT actuator bonded on the center of the plate with excitation frequency 650 rads/s for three different sizes of actuators.

5.2. Variation of Piezoelectric Actuator Locations

In this example, the piezoelectric actuators are surface bonded at two different locations, right and top region of the plate, respectively, as shown in Figure 14. These two typical locations were arbitrarily chosen to demonstrate the influence of the actuator location on the vibration shape. The piezoelectric actuators are excited by time harmonic voltages with three different frequencies of 350, 870 and 1,050 rads/s, respectively.

5.2.1. PZT Actuator Bonded on the Top Region of the Composite Plate

The vibration profile of the composite plate excited by the PZT actuators surface bonded on the top region of the plate with excitation frequency 870 rads/s are shown in Figure 15. The displacement amplitudes of the plate along the horizontal line $y = b/2$ and vertical line $x = a/2$ are presented in Figure 16. The vibration displacements of the plate obtained by the present approach of Equation (12) and finite element method are in close agreement. Table 7 lists the maximum amplitudes of the plate induced by the piezoelectric actuators surface bonding on the top location. It shows that the difference between the present approach and finite element method is within 8%.

Table 6. Maximum vibration amplitude (mm) excited by three different sizes of PZT actuator bonded on the center of the plate with excitation frequency 650 rads/s.

PZT Size	Finite Element ANSYS	Theoretical Prediction Equation (12)	Difference %
0.06 m × 0.04 m	0.516×10^{-3}	$0.508 \times 10^{-}3$	1.5
0.08 m × 0.06 m	0.100×10^{-2}	0.980×10^{-2}	2.0
0.10 m × 0.08 m	0.158×10^{-2}	0.151×10^{-2}	4.7

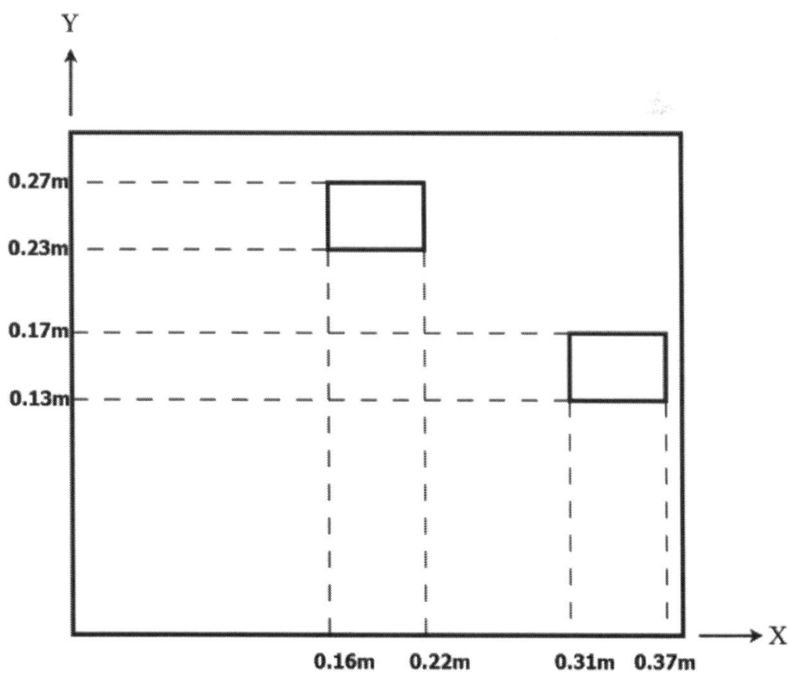

Figure 14. Two different locations of the PZT actuator.

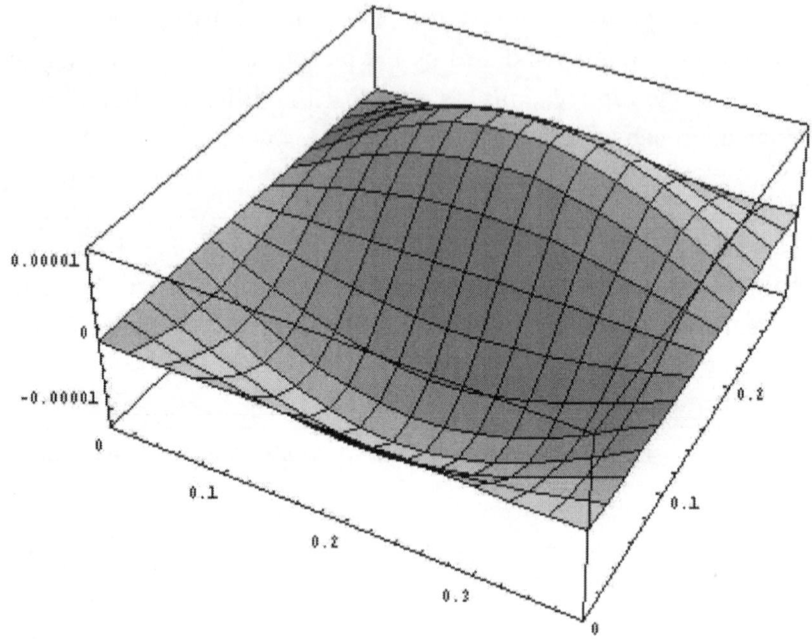

Figure 15. Vibration profile of the composite plate excited by PZT actuator bonded on the top region of the plate with excitation frequency 870 rads/s.

5.2.2. PZT Actuator Bonded on the Right Region of the Composite Plate

The vibration profile of the composite plate excited by the PZT actuators surface bonded on the right region of the plate with excitation frequency 1,050 rads/s are shown in Figure 17. The displacement amplitudes of the plate along the horizontal line $y = b/2$ and vertical line $x = a/2$ are presented in Figure 18. The vibration displacements of the plate obtained by the present approach of Equation (12) and finite element method are in close agreement.

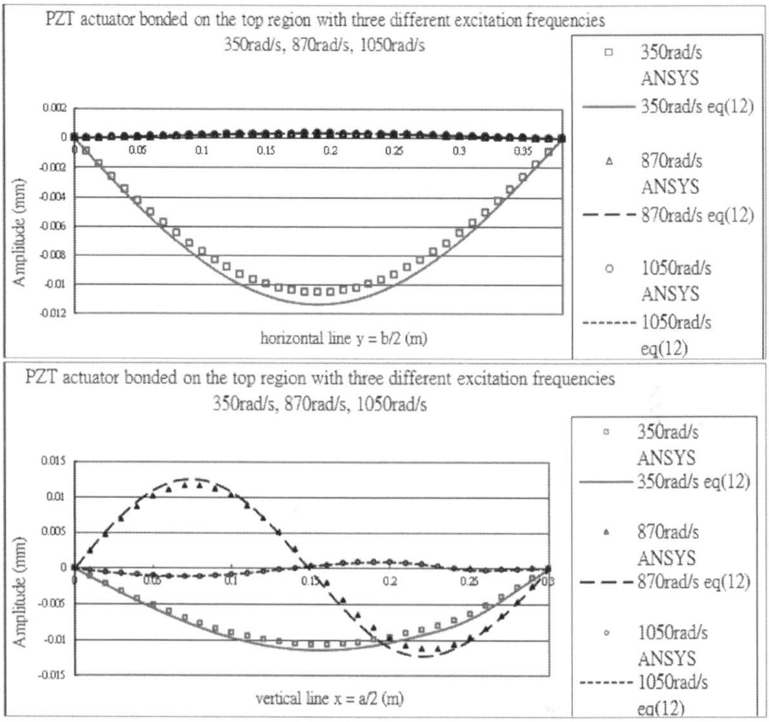

Figure 16. Vibration displacements of the composite plate obtained by ANSYS andEquation (12) along the horizontal line ($y = b/2$) and vertical line ($x = a/2$) excited by PZT actuator bonded on the top of the plate for three different excitation frequencies.

Table 7. Maximum vibration amplitudes (mm) excited by PZT actuator bonded on the top of the composite plate for three different excitation frequencies.

Frequency rads/s	Finite Element ANSYS	Theoretical Prediction Equation (12)	Difference %
350	1.05×10^{-2}	1.13×10^{-2}	7.4
870	1.16×10^{-2}	1.25×10^{-2}	7.3
1050	1.11×10^{-3}	1.05×10^{-3}	4.8

Table 8 lists the maximum amplitudes of the plate induced by the piezoelectric actuators surface bonding on the right region. It shows that the difference between the present approach and finite element method is within 7%. The difference of the vibrating shapes shown in Figures 9, 15 and 17 demonstrates that the vibration of the plate can be controlled by placing the actuators at various locations.

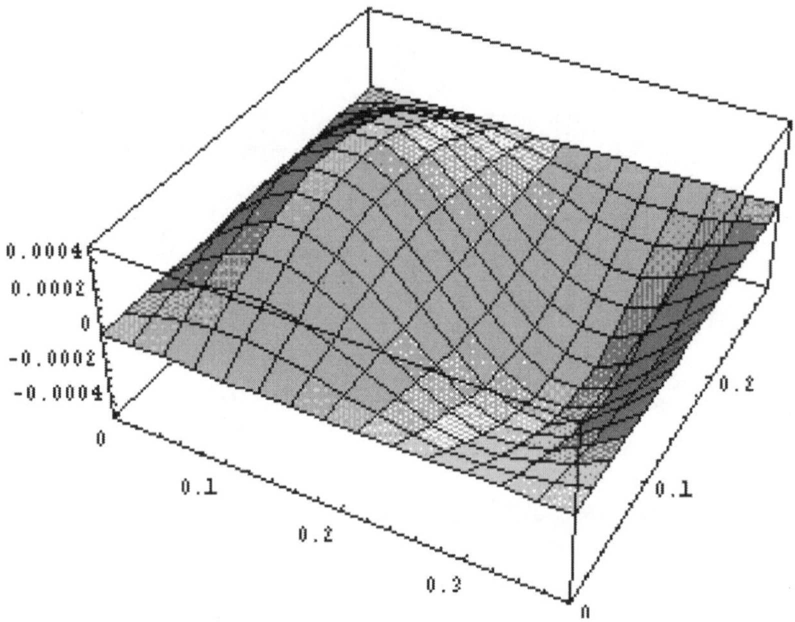

Figure 17. Vibration profile of the composite plate induced by PZT actuator bonded on the right region of the plate with excitation frequency 1,050 rads/s.

6. CONCLUSIONS

A model for the excitation of a composite laminate plate by patch type piezoelectric actuators surface bonded to the structure has been developed. The model has been applied to a simply supported cross-ply composite laminate plate excited by two piezoelectric actuators symmetrically bonded to both sides of the plate with time harmonic electric loading. An analytical expression of the harmonic vibration of the simply supported composite laminate induced by the piezoelectric actuators is presented. Numerical examples demonstrate that it is possible to excite the vibration modes by patch type actuators. When the input frequency is close to the resonant frequency of a mode, that mode can be excited provided the actuator is properly located. Three-dimensional finite element analysis was conducted using the commercial software ANSYS, and compared with the theoretical prediction. Good agreement was observed between the numerical result and theoretical prediction. Parametric study shows that the input frequency and location significantly affect the ability of the piezoelectric actuators to excite certain modes. Present study demonstrates the potential of controlling vibration in a plate structure using two dimensional patch type actuators. The optimal placement and configuration of PZT actuators will be the subject of future work.

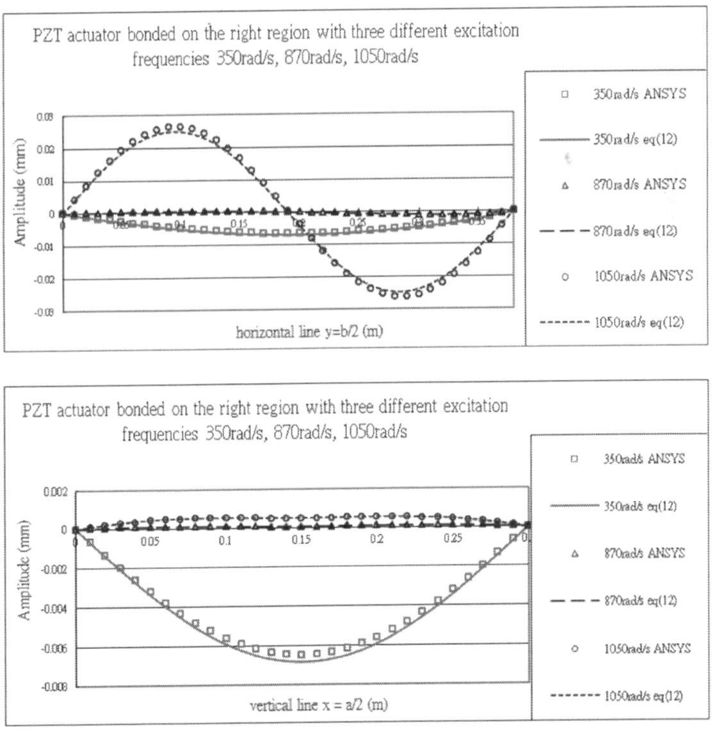

Figure 18. Vibration displacements of the composite plate obtained by ANSYS andEquation (12) along the horizontal line ($y = b/2$) and vertical line ($x = a/2$) excited by PZT actuator bonded on the right region of the plate for three different excitation frequencies.

Table 8. Maximum vibration amplitudes (mm) excited by PZT actuator bonded on the right region of the composite plate for three different excitation frequencies.

Frequency rads/s	Finite Element ANSYS	Theoretical Prediction Equation (12)	Difference %
350	6.48×10^{-3}	6.84×10^{-3}	5.2
870	1.09×10^{-3}	1.01×10^{-3}	7.0
1050	2.64×10^{-3}	2.49×10^{-2}	6.0

ACKNOWLEDGMENTS

The authors thank for the finical support by National Science Council; Taiwan; under the grand NSC 101-2221-E-155-013.

REFERENCES

1. Crawley, E.F.; De, L.J. Use of piezoelectric actuators as elements of intelligent structures. *AIAA J.* **1987**, *25*, 1373–1385.

2. Izadi, M.; Tahani, M. Analysis of interlaminar stresses in general cross-ply laminates with distributed piezoelectric actuators. *Compos. Struct.* **2010**, *92*, 757–768.

3. Kwak, M.K.; Heo, S.; Jeong, M. Dynamic modeling and active vibration controller design for a cylindrical shell equipped with piezoelectric sensors and actuators. *J. Sound Vib.* **2009**, *321*, 510–524.

4. St-Amant, Y.; Cheng, L. Simulations and experiments on active vibration control of a plate with integrated piezoceramics. *Thin Wall. Struct.* **2000**, *38*, 105–123.

5. Alibeigloo, A.; Madoliat, R. Static analysis of cross-ply laminated plates with integrated surface piezoelectric layers using differential quadrature. *Compos. Struct.* **2009**, *88*, 342–353.

6. Proulx, B.; Cheng, L. Dynamic analysis of piezoceramic actuation effects on plate vibrations. *Thin Wall. Struct.* **2000**,*37*, 147–162.

7. Susanto, K. Vibration analysis of piezoelectric laminated slightly curved beams using distributed transfer function method. *Int. J. Solids Struct.* **2009**, *46*, 1564–1573.

8. Qiu, Z.; Zhang, X.; Wu, H.; Zhang, H. Optimal placement and active vibration control for piezoelectric smart flexible cantilever plate. *J. Sound Vib.* **2007**, *301*, 521–543.

9. Santos, H.; Mota, S.C.M.; Mota, S.C.A.; Reddy, J.N. A finite element model for the analysis of 3D axisymmetric laminated shells with piezoelectric sensors and actuators. *Compos. Struct.* **2006**, *75*, 170–178.

10. Fernandes, A.; Pouget, J. Structural response of composite plates equipped with piezoelectric actuators. *Comput. Struct.* **2006**, *84*, 1459–1470.

11. Qing, G.; Qiu, J.; Liu, Y. A semi-analytical solution for static and dynamic analysis of plates with piezoelectric patches.*Int. J. Solids Struct.* **2006**, *43*, 1388–1403.

12. Huang, G.H.; Sun, C.T. The dynamic behaviour of a piezoelectric actuator bonded to an anisotropic elastic medium.*Int. J. Solids Struct.* **2006**, *43*, 1291–1307.

13. Dimitriadis, E.K.; Fuller, C.R.; Rogers, C.A. Piezoelectric actuators for distributed vibration excitation of thin plates. *J. Vibr. Acoust.* **1991**, *113*, 100–107.

14. Qu, G.M.; Li, Y.Y.; Cheng, L.; Wang, B. Vibration analysis of a piezoelectric composite plate with cracks. *Compos. Struct.* **2006**, *72*, 111–118.

15. Della, C.N.; Shu, D. Vibration of beams with piezoelectric inclusions. *Int. J. Solids Struct.* **2007**, *44*, 2509–2522.

16. Her, S.C.; Lin, C.S. Deflection of cross-ply composite laminates induced by piezoelectric actuators. *Sensors* **2010**, *10*, 719–733.

17. Patnaik, B.R.; Heppler, G.R.; Wilson, W.J. Sensor Effectiveness Coefficients for Piezoelectric Materials. Proceedings of the American Control Conferences, Seattle, WA, USA, 30 June–5 July 1996; pp. 3801–3802.

CHAPTER 4

Thermal Activation of Asymetrical Composites for Vibration Control

Axel Imbert[1,2], Gildas L'Hostis[1*], David Rigel[2], Fabrice Laurent[1], Bernard Durand[1]

[1] Laboratoire de Physiques et Mécanique, Ecole Nationale d'Ingénieur de Sud Alsace, Mulhouse, France

[2] Centre Technique des Industrie Mécanique, Centre Recherche Matériaux Textiles, Mulhouse, France

ABSTRACT

The CBCM (Controlled Behaviour Composite Material) is a thermal active composite, which has been developed for morphing applications. The thermal activation is made by a source of heating generated within the composite structure. The coupling between the induced thermal field and the thermomechanical properties of the various components of the composite structure leads to the change of the structure shape. The heat source is generated by Joule effect, Carbon yarns inserted in the composite, are connected to a power supply. The application field of CBCM technology is the domain of shape modification and active assembly. The objective of this work is to illustrate the capabilities of CBCM in the domain of vibration control. We will study several reference plates with different constitution. The influences of these different constitutions, of the CBCM effect and the loss of stiffness for the matrix will be highlighted, for two boundary conditions, free/free and embedded/embedded.

KEYWORDS

Composite Material; Vibration Control; Smart Materials; Active Structure

1. INTRODUCTION

Smart materials can be defined as materials which are able to change their form, their mechanical or any physical properties to adapt to their background. They are more and more used for technical applications and particularly in the field of controlling structural deformations. In this case, main actuators use piezoelectric and shape memory alloys materials. To make an active composite, these actuators may be used at the scale of the fiber or the matrix [1,2], at the scale of the interface between the successive layers in a laminate [3,4] or at the scale of the linkage between the actuator and a passive composite structure [5,6]. However, the interface strength between the actuator and the composite plays a crucial role. It is a limit of such a technology for the time life of the structure, especially when the rigidity of the composite structure is high, because to deform the structure the actuator has to fight against this rigidity. The use of a thermal actuation, based on the bimetallic strip effect can be a solution to overcome the problems of bonding between the actuator and the structure. Indeed, the whole structure can be considered as an actuator and the problem of interface cohesion is not concentrated at the interface actuator/structure but distributed all along the interfaces of the laminated composite. For composite structures, the bimetallic strip effect can be obtained by coupling an internal heat source with an unsymmetrical laminate.

The concept of a thermal activation based on a bimetallic effect, CBCM effect [7,8] has been studied for morphing structure, the objective of this work is to show the ability of this technology in the field of vibration control of composite structures. For the control of vibration the main actuator remains the piezoelectric, for example the Macro Fiber Composite [9,10]. However in the literature, some application using SMA actuators can be found. The particular behavior of the SMA actuators inserted in the composite or fixed on the composite surface, leads to two technics of vibration control which lead to a frequency shift for different eigenvalue. The ASET technic (Active Strain Energy Tuning) [11], use the shape memory property of the SMA. A non-activated shape and an activated shape are programmed for the SMA actuator by a thermo mechanical cycle. SMA actuator is inserted in the composite have to be maintained during the manufacturing of the composite. During actuation the shape recovery of the SMA induces a stress field and modifies the dynamic behavior of the composite. To modify the dynamic behavior, the APT technic [12,13] uses the change of the SMA behavior during the phase transformation (variation of the Young's modulus between austenitic and martensitic phase.

A CBCM combines these two actions, stress field increase and change for the behavior of the matrix, especially a loss of stiffness during heating.

After a presentation of the CBCM principle, the constitution of the different tested plates, the experimental procedure, boundary conditions and experimental apparatus will be totally described. In a first par for two kinds of boundary condition, free/free and embedded/embedded and for two CBCM, the influence of the plate rigidity on their dynamical response will be studied: for nonactivated and activated plates. The shift of frequency for the different eigenvalue will be highlighted. In the second part of this work, the influence of the loss of stiffness for the matrix will be studied for a

CBCM plate and a symmetrical plate. An embedded/embedded boundary condition will be use and from the experimental date, the shape of the two first modes will be reconstructed.

2. MATERIAL AND METHODS

2.1. How CBCM Works

The bending or the warping is a well-known drawback of a dissymmetrical composite when submitted to temperature variations.

Two different manners to have a bending effect by heating the CBCM have been identified: one called “single effect” or “temperature effect”, another called “double effect” or “gradient effect”. The characterization between “single” or “double” is a reference to the possibility for CBCM plates to bend in one or two directions. The single effect is obtained thanks to an asymmetric laminate made of layers with different coefficients of thermal elongation in a given direction. The different coefficients can be obtained either by different materials or by different orientations of the reinforcement in the layers.

The double effect can be obtained with any composite structure containing an insulating layer, for instance a sandwich structure. If only one side of the composite is heated, the other side being insulated, a gradient of temperature appears within the structure. The result is, as previously shown, the bending of the plate **Figure 1**.

The thermal response time of the CBCM and conesquently the response time of the structure is the same order as the SMA one [14]. This response time is directly linked to the intensity of the current in the active layer.

2.2. Description of the CBCM Plates

The composite plates (395 × 125 × 2 mm^3), prepared by compression moulding, are made of seven layers: two layers ($2D^G$) of plain weave Glass fabric (196 g/m^2), two layers (90^G) of Glass unidirectional (588 g/m^2), an active layer (A_L) and two layers ($2D^A$) of plain weave Aramid fabric (173 g/m^2). Three constitutions of plates are studied: two CBCM plates SL = $\{(2D^G)_2/(0^G)_2/A_L/(2D^A)_2\}$ and ST = $\{(2D^G)_2/(90^G)_2/A_L/(2D^A)_2\}$where 0° is along the longitudinal direction of the plate. The difference between the two CBCM plates becomes to the orientation of the main reinforcement, the glass unidirectional. The latest constitution is symmetrical SYM = $\{(2D^G)_1/(90^G)_1/(2D^A)_1/AL/(2D^A)_1/(90^G)_1/(2D^G)_1\}$. The active layer is made of four parallel carbon yarns (6000 filament of 7 μm diameter) along the longitudinal direction of the plate. The yarns are connected to a DC generator and provide the internal heat source of the structure. The resin used for the matrix is an epoxy resin (Epolam 20 - 25 from Axson Corporation). The curing cycle followed is recommended by the manufacturer to obtain a glass transition temperature Tg of 130°C.

2.3. Experimental Procedure

In this work two experimental procedures are used. The first procedure consists to characterize the influence of the internal stress field induced by CBCM effect on the whole dynamic response of plates with different rigidity (SL and ST plates). Two kinds of boundary conditions are considered: free/free and embedded/embedded. To perform the dynamic analysis, the measuring systems is composed by a Brüel & Kjær acquisition system, an impact hammer (stress sensor PCB 086C05 # 13381), a piezoelectric accelerometer (Brüel & Kjaer 4393# 1803841, sensitivity 0319 pC/ms^{-2}), an amplifier (Endevco 133 # BD69, gain 10 mV/ms^{-2}) and a Pulse acquisition module with I/O 4/2 way 3109. For the value of a modal frequency, this set gives accuracy equal to ±2 Hz.

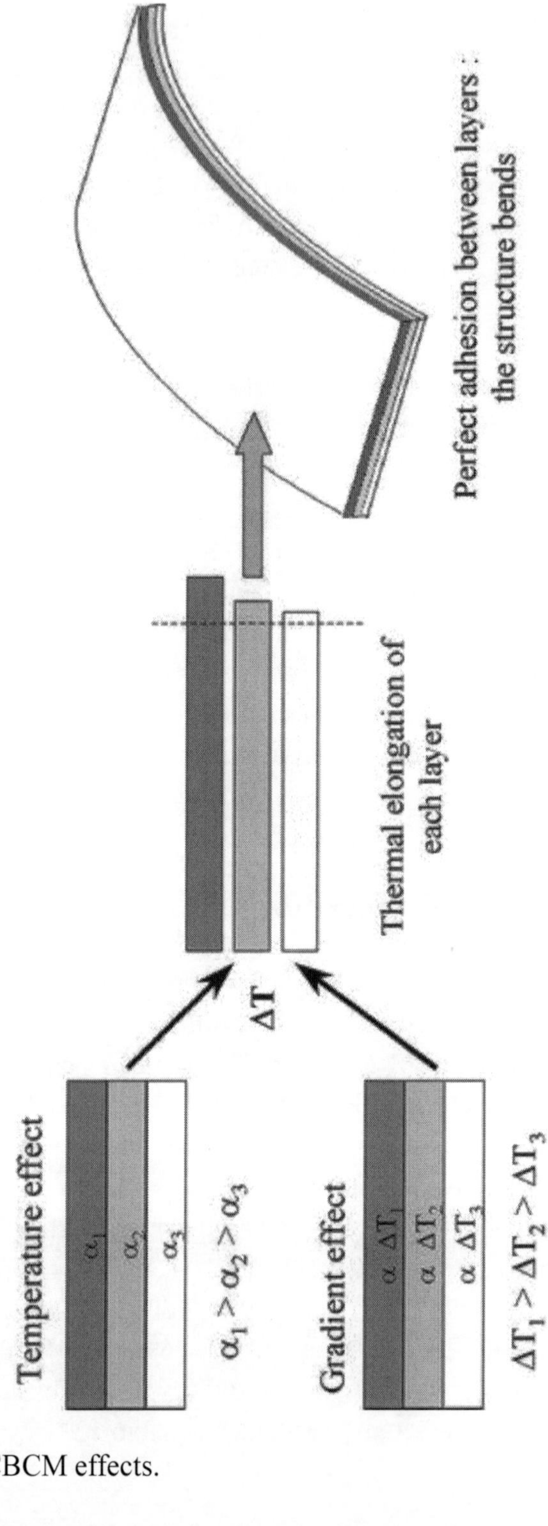

Figure 1. CBCM effects.

The second experimental procedure consists to highlight the influence of change of rigidity of the plates due to the mechanical softening of the matrix function due to the temperature increase. The plate is embedded on line A and B **Figure 2**. The measurement area (240 × 125 mm^2) is discretized in 20 measurement nodes. The dynamic analysis is performed using a system acquisition, Brüel & Kjaer, an impact hammer (with force sensor PCB 086C05 # 13381), a front acquisition Pulse (Controller Module 7539 A). The speed of the plate is measured using a laser vibroscope (Ometron DV-500-D) in the node 1 as the reference node. During the test, the 20 nodes are successively affected by the hammer. The modal frequency accuracy is equal to 0.2 Hz. After treatment, in order to construct the shape of the different modes, experimental data are transferred to ME'scope VES software.

The temperature of the plate is measured by thermocouple on the upper side of the plate and a DAQ NX acquisition system of National Instrument.

3. RESULTS AND DISCUSSION

3.1. Influence of the Plate Rigidity

The difference of rigidities between SL and ST plates is due to the orientation of the two glass unidirectional layers. The longitudinal rigidity is greater for the SL plate than the ST one; which is the opposite for the transversal rigidity.

For the free/free boundary conditions, **Figure 3** shows the responses of the two plates at ambient temperature equal to 20°C (without the activation of CBCM effect). The smaller longitudinal rigidity of the ST plate leads to a decreasing of the modal frequency. This result is in accordance with those obtained during the static characterization of plates for a free point bending test. For the SL and ST plates, the flexural modulus is 20 GPA and 16 GPA respectively. It can be noted that the morphing properties free displacement and blocking force given by the unconstrained and constrained test respectively are higher for the ST plate [14].

For a temperature of activation $T_A = 100°C$, **Table 1** summarizes the results obtained for the first seven modes of the plates. δ is the difference (in %) between the frequency modal values of the non-activated and activated plates. the non-indicated values correspond to a lower value than the accuracy of the measurement system (4 Hz). For the two plates, the CBCM effect affects the same values of modal frequency. The maximal values of δ are observed for the ST plate. For mode 1 and 3 no change of frequency may be observed.

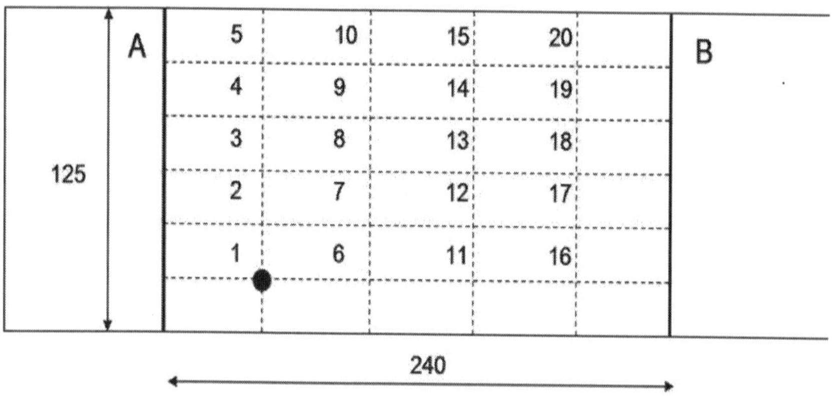

Figure 2. Discretized plate.

For the free/free boundary conditions and due to the morphing property of the plates, the direction of curvature changes between the non-activated and activated plate **Figure 4**.

For the embedded/embedded test the change of curvature is cancelled by the boundary condition. At ambient temperature, the values of modal frequency **Table 2** show the difference of dynamical response of the two plates. Contrary to the precedent results, the ST plate is more rigid than the SL one. This change of rigidity becomes to the boundary conditions, because due to the clamping an internal stress field appears in the plates, higher in the ST plates due to the higher value of the initial curvature.

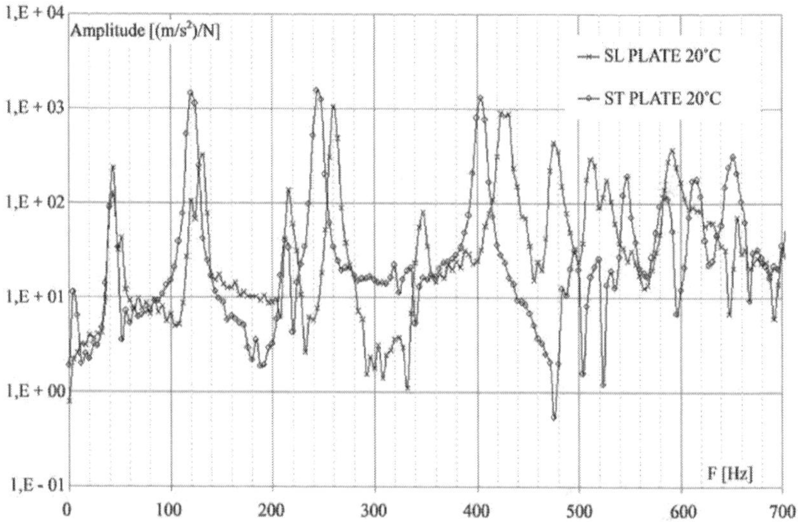

Figure 3. SL and ST plates dynamical response at 20°C (free/free boundary conditions).

For T_A = 100°C the CBCM effect can be observed for the SL plate **Figure 5** and the ST plate**Figure 6**. The same kinds of previously results are obtained: no changes for certain frequency modal value and for the modes a lower value for the activated plates. After activation, the rigidity of the ST plate remains higher than the SL plate (**Table 2**), however the maximum values of δ are obtained for the ST plate. Depending of the glass unidirectional orientation, the CBCM effect not affects the same nodal frequency values, and the influence of the CBCM effect is different depending on the frequency. The first three modes are unchanged for the SL plate; the third mode of the ST plate has the maximum value of δ. An opposite result is obtained for the fourth mode of the plate SL. This difference of dynamical behavior between SL and ST plates may be induced by the initial stress field due to the embedded boundary conditions. But the thermal activation has an influence on the mechanical property of the matrix and induces a loss of stiffness.

Table 1. Frequency modal value for free-free boundary conditions (Hz).

	SL			ST		
Mode	20°C	100°C	δ (%)	20°C	100°C	δ (%)
1	44	44	-	4	6	-
2	120	104	13	44	36	18
3	132	128	-	120	120	-
4	216	192	11	212	180	15
5	260	232	11	244	236	3
6	348	312	10	332	296	11
7	428	416	3	404	388	4

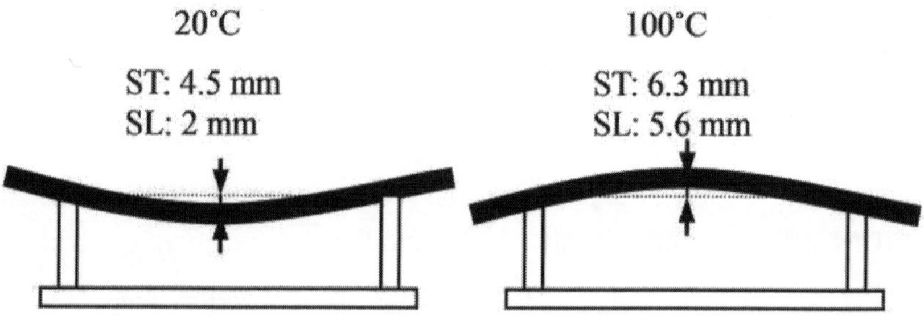

Figure 4. Curvature change for SL and ST plates.

Table 2. Frequency modal value for embedded/embedded boundary conditions (Hz).

	SL			ST		
Mode	20°C	100°C	δ (%)	20°C	100°C	δ (%)
1	32	32	-	32	32	-
2	60	60	-	92	92	-
3	108	104	-	128	112	13
4	148	136	8	252	248	-
5	292	284	3	304	276	9
6	496	472	5	492	480	2
7	648	604	7	660	604	8

Thereby, when the plate is heating three phenomena are in competition. The initial stress field and the stress field induce by the CBCM effect. Linking to the asymmetry of the composite, these fields are higher for the ST plate and lead to an increase in stiffness along the longitudinal direction of the plate. The third phenomenon is the loss of stiffness and for the ST plate, due to the orientation of the glass unidirectional; this phenomenon's effect is higher for the longitudinal direction.

3.2. Influence of the Matrix Loss Stiffness

For the embedded/embedded boundary conditions, the dynamic behavior of the thermal active plates is influenced by three coupled phenomena. The initial and thermal stress fields due to the curvature of the plate and the CBCM effect respectively, lead to an increase of the stiffness. On the opposite, the latest phenomenon induces a loss of rigidity linking to the matrix loss stiffness. To isolate this phenomenon, two plates are studied the ST plate and a symmetrical plate SYM. The dynamic analysis is made of two steps. The first is the determination of the modal frequency but only for the two first modes. The second step consists to reconstruct the associated modal shape of the modes.

For node 1, the reference point **Figure 2**, a change of modal frequency is observed for the SYM plate **Figure 7**. The value of δ between the plate at T_{amb} and the activated plate at T_A = 100°C is δ = 12% for the first mode (10.3 Hz), and δ = 25% for the second mode (42 Hz). The corresponding damping coefficients are respectively, 1.72% and 1.15% for the non-activated plate and 3.22% and 3.35% for the activated plate.

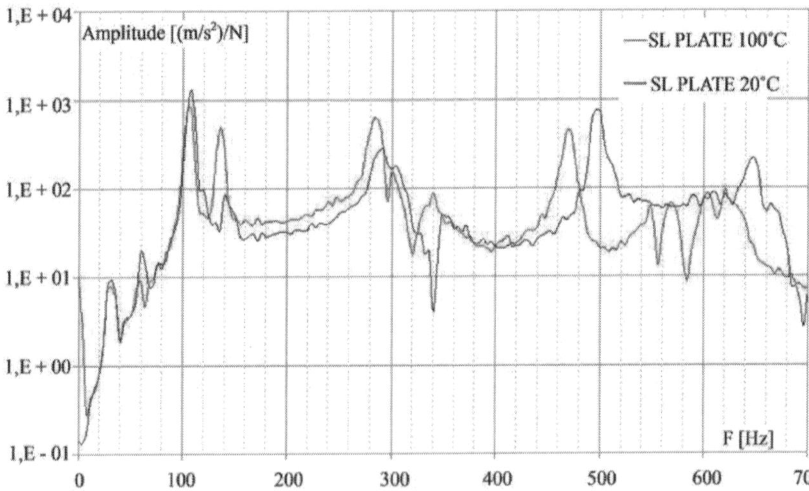

Figure 5. SL plate dynamical response (embedded/embedded boundary conditions).

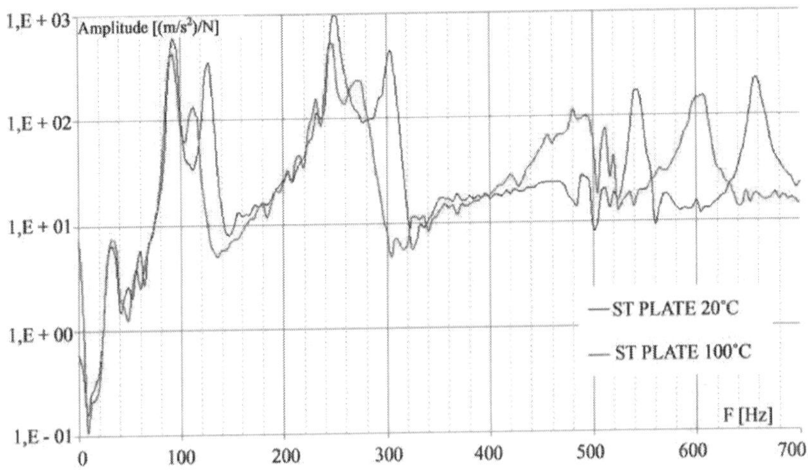

Figure 6. ST plate dynamical response (embedded/embedded boundary conditions).

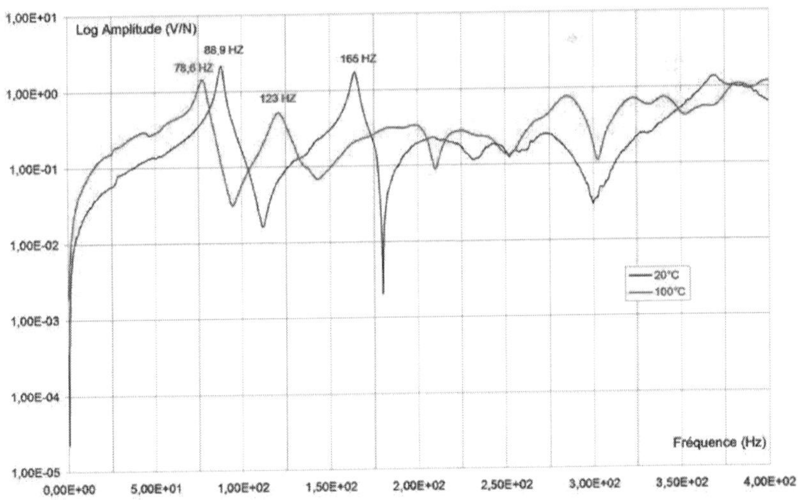

Figure 7. Dynamic response, SYM plate, embedded/embedded, at 20°C and 100°C.

The previous results obtained for the ST plate are confirmed **Figure 8**. The value of δ between the plate at T_{amb} and at T_A = 100°C is δ =

3% for the first mode (4 Hz) and δ =12% for the second mode (20 Hz). The corresponding damping coefficients are respectively, 1.35% and 1.14% for the non-activated plate and 2.48% and 2.39% for the activated plate.

For the two plates, the thermal activation leads to a decrease in the value of the frequency, and an increase of the damping coefficients. This result is consistent with the variation of the behavior of the epoxy depending to the temperature increase: loss of rigidity and increase of viscous properties.

However, for the asymmetrical plate ST, the changes of modal frequency and damping factor depending to the temperature are smaller than for the symmetrical plate. For the symmetrical plate, the changes are only due to the loss of stiffness of the epoxy resin depending to the temperature. For the ST plate, during the thermal activation, the competition between CBCM effect, and the loss of epoxy stiffness remains. For the two modes of the ST plate, the coupling between these opposite phenomena acts differently. For the first mode, the loss of stiffness is partly compensated by the CBCM effect.

This result may explain the dynamical behavior observed in Section 3.1 (Tables 1 and 2). For certain mode the loss of stiffness is totally compensated by the CBCM effect, and the value of the modal frequency remains the same. For the other modes the loss of stiffness is not compensated and a change of frequency is obtained. So depending on the shape of the mode, the coupling acts differently.

From experimental data, the shape of the two modes is reconstructed. For the SYM plate **Figure 9**, at ambient temperature the two modes are for the first a (2,1) mode and a (1,2) mode for the second. After the thermal activation, the same modal shapes are observed. The mode most affected by the loss stiffness (δ = 25%) is the second mode (1,2). This result is consistent with the orientation of the main reinforcement, the Glass unidirectional, along the transversal direction of the plate.

For ST plate **Figure 10**, the shape of the modes is conserved between the non-activated and activated structure. The modes are: (1,1) mode for the first and (1,2) for the second. As previously for the SYM plate, the second mode, the same that the SYM plate, is more affected by the loss stiffness ($\delta = 12\%$). However the first mode remains quasi constant ($\delta = 3\%$). The bending effect along the longitudinal direction of the plate due to CBCM effect affect mainly this first mode composed by a flexural mode along the longitudinal direction. Along this direction, the CBCM effect balances the loss of stiffness.

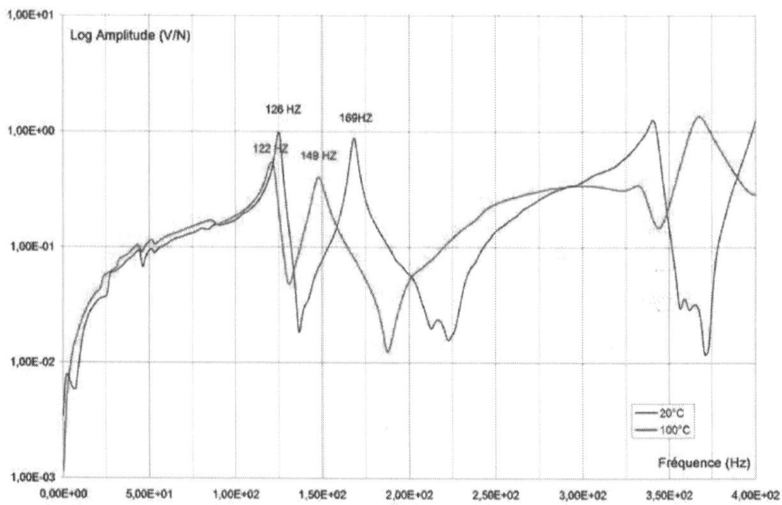

Figure 8. Dynamic response, ST plate, embedded/embedded/ at 20°C and 100°C.

4. CONCLUSIONS

For the all modes, the softening of the matrix depending of the temperature increase induces an important change for their modal frequency. This result has been also observed for inserted SMA, for their temperature of activation. The loss of rigidity of the matrix is accompanied by an increase in its viscous properties. For a SYM plate, a thermal active layer may be used to change its dynamic behavior, but all modes will be modified by this thermal activation. For the asymmetrical plate, the CBCM effect is added and for certain modes can be compensated the softening effect.

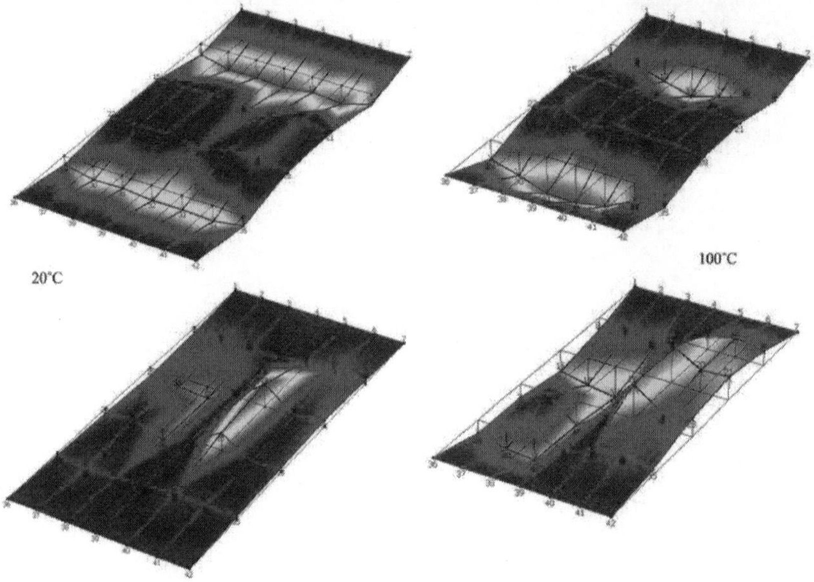

Figure 9. Modal shape, SYM plate.

The softening is induced by the behavior of the matrix and is linked to the heat field in the composite plate. Depending to this thermal field, the softening can be local or global at the entire plate. The CBCM effect is also linked to the heat field; it can be local or global. However the CBCM effect depends also to the organization of the structure nature and orientation of the reinforcement. Therefore the coupling depending to the thermal field, between the loss of rigidity and the CBCM effect gives a wide field of possibility for the adaptation of the composite structures dynamic behavior.

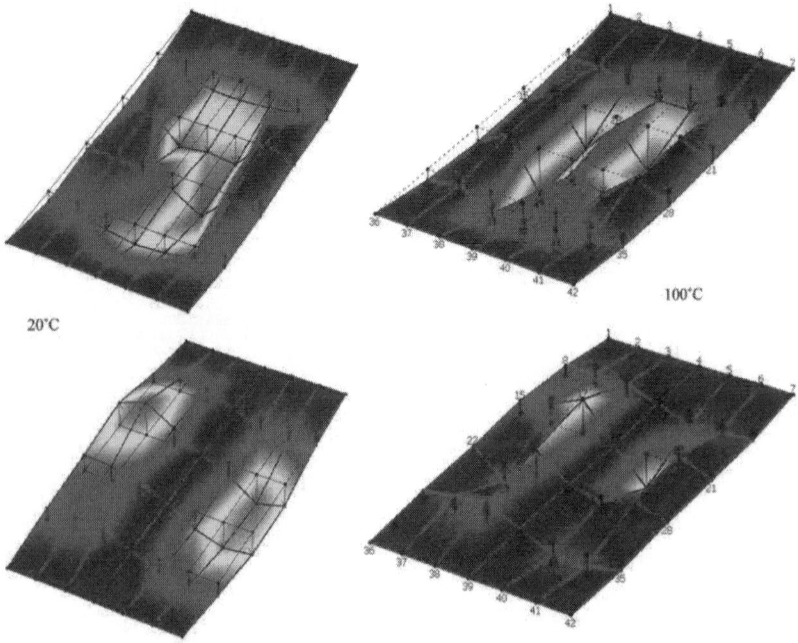

Figure 10. Modal shape, ST plate.

Finally, one of the main advantages of the thermal activation compared to SMA or piezoelectric technologies is its simplicity of implementation in the process of composite structures manufacturing and this for any kinds of structure geometry.

REFERENCES

1. G. C. Psarra, J. Parthenios and C. Galiotis, "Adaptative Composites Incororating Shape Memory Alloy Wires," Journal of Material Science, Vol. 36, No. 3, 2001, pp. 535-546.

2. M. V. Donadon, S. F. M. Almeida and A. R. de Faria, "Stiffening Effects on the Natural Frequencies of Laminated Plates with

Piezoelectric Actuators," Composite Part B, Vol. 33, No. 5, 2002, pp. 335-342.

3. C. S. Zhang and Q. Q. Ni, "Bending Behaviour of Shape Memory Polymer Based Laminates," Composites Structures, Vol. 78, No. 2, 2005, pp. 153-161.
4. D. M. Elzey, A. Y. N. Sofla and H. N. G. Wadley, "A Shape Memory-Based Multifunctional Structural Actuator Panel," International Journal of solids and Structures, Vol. 42, No. 7, 2005, pp. 1943-1955.
5. K. J. Yoon, K. H. Park, H. C. Park and D. Perreux, "Thermal Deformation Analysis of Curved Actuator LIPCA with a Piezoelectric Ceramic Layer and Fiber Composite Layers," Composite Science and Technology, Vol. 63, No. 3-4, 2003, pp. 501-506.
6. M. R. Schultz, M. W. Hyer, R. B. Williams, W. K. Wilkie and D. J. Inmann, "Snap-Through of Unsymmetric Laminates Using Piezocomposite Actuators," Composites Science and Technology, Vol. 66, No. 14, 2006, pp. 2442- 2448.
7. H. Asanuma, O. Haga, J. I. Ohira, G. Hakoda and K. Kimura, "Proposal of an Active Composite with Embedded Sensor," Science and Technology of Advanced Material, Vol. 3, No. 2, 2002, pp. 209-216.
8. H. Drobez, G. L'Hostis, K. Gautier, F. Laurent and B. Durand, "A New Active Composite," Journal of Smart Material and Structures, Vol. 18, No. 2, 2009, Article ID: 025020.
9. A. Kovalovs E. Barkanov and S. Gluhihs, "Active Control of Structures Using Macro-Fiber Composite (MFC)," Journal of Physics: Conference Series, Vol. 93, No. 1, 2007, Article ID: 012034.

10. G. V. Kumar, S. Raja, K. B. Prasanna and V. Sudha, "Finite Element Analysis and Vibration Control of a Deep Composite Cylindrical Shell Using MFC Actuators," Smart Materials Research, Vol. 2012, 2012, Article ID: 513271.

11. S. John and M. Hariri, "Effect of Shape Memory Alloy Actuation on the Dynamic Response of Polymeric Composite Plates," Composites Part A, Vol. 39, No. 5, 2008, pp. 769-776.

12. K. Yuse and Y. Kikushima, "Development and Experimental Consideration of SMA/CFRP Actuator for Vibration Control," Sensors and Actuators A, Vol. 122, No. 1, 2005, pp. 99-107.

13. R. Zhang, Q. Ni, A. Masuda, T. Yamamura and M. Iwamoto, "Vibration Characteristics of Laminated Composites Plates with Embedded Shape Memory Alloys," Composite Structures, Vol. 74, No. 4, 2006, pp. 389-398.

14. K. Gautier, G. L'Hostis, F. Laurent and B. Durand, "Mechanical Performances of a Thermal Activated Composite," Composite Science and Technology, Vol. 69, No. 15- 16, 2009, pp. 2633-2639.

CHAPTER 5

Vibration Control for Two air Compressor units Used in Tandem

Viorel-Mihai Nani[1,2], Ioan Cireş[2]

[1] Faculty of Engineering, University "Ioan Slavici" Timisoara, Timisoara, Romania

[2] Research Department, SC PRO ATLAS ING SRL Timisoara, Timisoara, Romania

ABSTRACT

The paper presents the vibration control for two air compressors used in tandem to an oxygen generating station. The compressors are identical in terms of constructive and functional proper- ties, and are both located on similar foundations. Each compressor is supported on a foundation made from one massive block of concrete, provided with some elastic and damping parts. This anti vibrating structure insulates the machines compared to other station equipment. During operation, were observed dangerous levels of vibration at one of the machines, which forced some un- wanted stoppages for the station. Two hypotheses have been promoted. One of them referred to a pronounced wear of one compressor, although this was relatively new. The second hypothesis has taken in consideration the appearance of some cracks in the foundation massive concrete block.

Experimental research conducted on the spot led to the identification of some errors done at foundation fabrication. The experimental results and some theoretical investigations are pre- sented in this paper.

KEYWORDS

foundation, harmonic disturbing Forces, Amortized forced Vibration, Vibration isolation, Coefficient of transmissibility

1. INTRODUCTION

In general, operating industrial equipments produce vibrations under the action of some disturbing forces. These are usually, periodic time dependent external forces. Common source for disturbing forces are the inertia forces. Virtually every moving body—(movements like: rotation, linear displacement, linear—alternative, oscillatory etc.)—is subject to inertial forces and therefore produces vibrations [1] and [2].

Vibrations are transmitted to the foundation on which is installed the equipment. From dynamic stability con- siderations, these vibrations should be isolated. To isolate the vibrations it means to identify some technical measures, so that the foundation does not vibrate when the industrial equipment creates these forces. So, is re- quired that the disturbing forces transmitted to the foundation to be zero or as low as possible with the help of some damping measures [3] .

In the paper we studied a real case that occurred to an oxygen station. The replacement of two air compressors with a new model concurrently with restoration the foundation had generated unforeseen situation. During commissioning tests, there were identified dangerous levels of vibration which led the temporary shutdown of the compressors, and thus the oxygen station. In this context, our research team was asked to correct the non- compliance.

2. THEORETICAL CONSIDERATIONS ON FORCED VIBRATION

A particular important case in practical applications is the situation where the damping forces are negligible and over the mechanical system is acting only the disturbing force $F(t)$. Being independent of system characteris- tics, the differential equation of motion is [4] and [5] :

$$m\ddot{x} + kx = F(t) \tag{1}$$

where m—the mass of the mechanical system (industrial equipment); k—the elastic constant of the mechani- cal system; x—coordinate of motion.

In practice, is often encountered situations where the disturbing force is harmonic. If the pulsation of this force is ω, then equation (1) is:

$$m\ddot{x} + kx = F_0 \sin\omega t \tag{2}$$

This equation has the particular solution

$$x_f = x_0 \sin\omega t \tag{3}$$

where x_f—the amplitude for the forced vibration in the mechanical system; x_0—the amplitude of the vibration; Substituting (3) in (2) and putting condition $\omega^2 \neq \omega_n^2 = \frac{k}{m}$, we obtain:

$$x_0 = \frac{Q_0}{k - m\omega^2} \tag{4}$$

With the notation $\frac{F_0}{k} = x_{st}$, where x_{st} is the static deformation of the vibrating system under the action of a disturbing force equal with the maximum value F_0, Equation (4) can be written in dimensionless [3] :

$$\frac{x_0}{x_{st}} = \frac{1}{1 - \left(\frac{\omega}{\omega_n}\right)^2} \tag{5}$$

We can observe that the amplitude of vibration x_0 depends by the pulsation ratio ω/ω_n. This dependence isshown in Figure 1 [6].

Note that for $\omega < \omega_n$, x_0 is positive and represents the forced vibration amplitude, being higher than x_{st}.

In this situation, the forced vibration x_0 and disturbing force $F_0 \sin\omega t$ are in phase.

For $\omega > \omega_n$, x_0 is negative. In this case the forced vibration and disturbing force are out of phase with π, or they are in opposition. The vibration amplitude is $|x_0|$, but for the vibration study we will suppress the negative sign and we will consider the dashed drawn curve (see Figure 1).

For $\omega < \sqrt{2}\omega_n$, i.e. for low frequency of disturbing forces, the forced vibration amplitude is greater than x_{st}.

For high frequencies $\omega > \sqrt{2}\omega_n$, the amplitude is less than x_{st} and tends to be zero when ω tends to be $+\infty$.

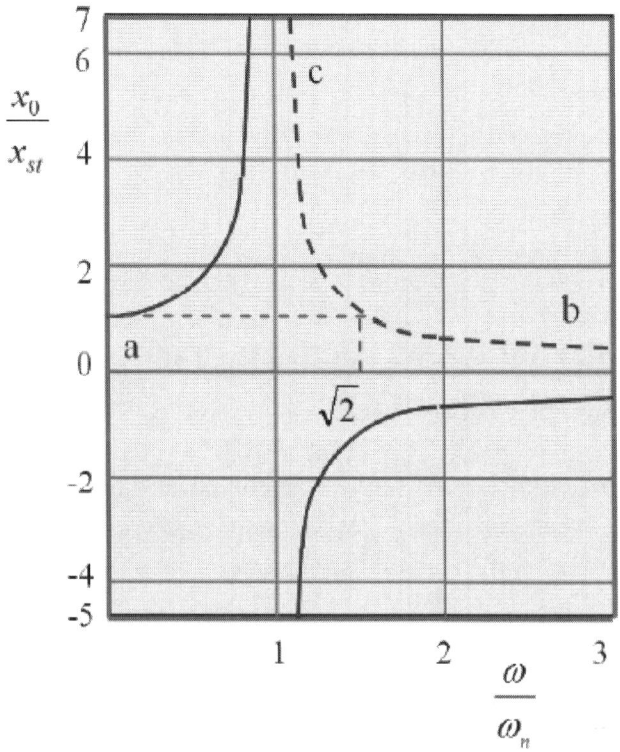

Figure 1. Diagram amplitudes depending pulses of the mecha- nical system.

In the point a, the disturbance force frequency is very low, $\omega \approx 0$ and the mechanical system vibrates in phase with the force with the amplitude equal to x_{st}. In the point b, $\omega \gg \omega_n$ the amplitude of forced vibration x_0 is very small. The force varies so rapidly that mechanical system does not have time to follow it. Basically, the system remains in relative rest. The point c, where $\omega = \omega_n$, corresponds to the phenomenon of resonance. The vibration amplitudes increase together with the forces of the mechanical system. If these forces exceed the limit values for the strength of the mechanical structure, the system can be destroyed irreversibly. For this reason, the design of industrial equipment and foundation, take into account the phenomenon of resonance.

Resonance is practically avoided either by changing frequency for disturbing force when it coincides with the resonance frequency, either by modifying the structure of the system, by modifying the parameters m and k. The second situation includes also the computation for the foundation isolation [1] and [7].

3. THEORETICAL CONSIDERATIONS ON VIBRATION ISOLATION

To reduce or eliminate the forces transmitted to the foundation, it is inserted between the mechanical system and foundation different elastic materials: metal coil springs, cork, rubber, polystyrene, expanded sheets etc. If we note with k the elastic constant of the materials inserted between the mechanical system and the foundation, and with m mass of the industrial equipments, then the mechanical model is presented in Figure 2 [3] .

Transmissibility coefficient η, namely ratio between the maximum of force transmitted with foundation F_{tr} and disturbing force $F_0 \sin\omega t$, is given by dimensionless relation (5). As seen fromFigure 1, in order as force transmitted to foundation to be as small as possible, the ratio ω/ω_n must be as high as possible. This is done basically by adopting constructive solutions with their own pulsation $\omega_n = \sqrt{\frac{k}{m}}$ as small as possible, (i.e. springs supple with elastic constant k small).

If industrial equipment is placed directly on the foundation without elastic elements, $k \to +\infty$, $\omega_n \to +\infty$ and $\omega \approx \omega_n$, so the seating basis is rigid, the transmissibility coefficient $\eta \approx 1$, and disturbing forces are trans- mitted entirely to the foundation.

For $\omega < \sqrt{2}\omega_n$, the transmissibility ratio is $\eta > 1$. In this situation, whether there are introduced elastic mate- rials, the results increase the forces transmitted to the foundation F_{tr}. The effect is more pronounced if the pul- sation ω is close to the resonant pulsation ω_n. Then transmissibility coefficient η becomes very large, and the forces transmitted to the foundation, are very dangerous. Therefore, in the foundation structure, vibration isolation measures are introduced, in addition to elastic materials and viscous damping elements. The differen- tial equation of vibration movement in this case is [2] :

$$m\ddot{x} + c\dot{x} + kx = F(t) \qquad (6)$$

where c is the damping coefficient.

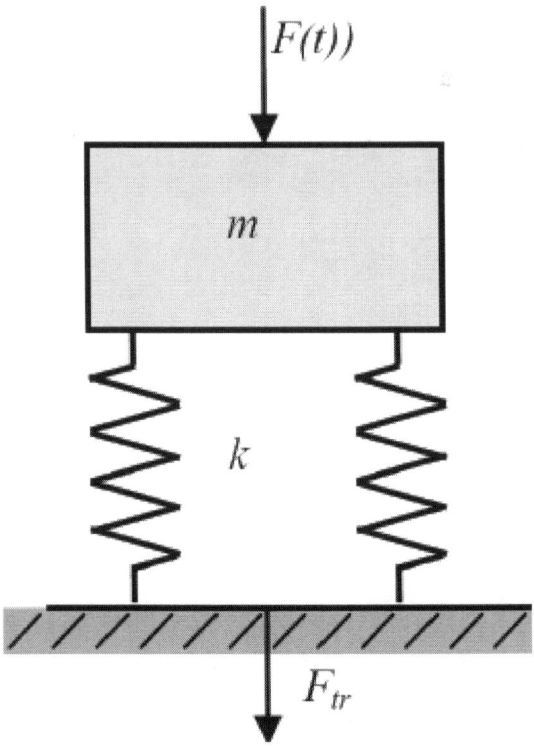

Figure 2. The mechanical vibration isolation with elastic elements.

If disturbing force is harmonic, namely $F(t) = F_0 \sin\omega t$, then forced vibrations x_f as solution of Equation (6) is:

$$x_f = x_0 \sin(\omega t - \varphi) \tag{7}$$

where φ is the initial phase at moment $t = 0$ and is obtained from

$$tg\varphi = \frac{c\omega}{k - m\omega^2} \tag{8}$$

At resonance, the maximum forced amplitude is given [6]:

$$x_{0\max} = \frac{x_{st}}{2\frac{c}{c_{cr}}\sqrt{1 - \left(\frac{c}{c_{cr}}\right)^2}} \tag{9}$$

where c_{cr} is the critical damping coefficient, being equal to:

$$c_{cr} = 2m\sqrt{\frac{k}{m}} = 2m\omega_n = 2\sqrt{mk} \tag{10}$$

Discussions

- In case $c > c_{cr}$, the movement is not vibratory, but atypical. Mechanical system slowly approach to the rest position $x = 0$ due the very high forces created by the insulation structure.

- Where $c = c_{cr}$, the mechanical system movement is also atypical and it slowly approach the rest position $x = 0$.

- If $c < c_{cr}$, mechanical system approaches rest position $x = 0$ in the time interval c/k called the relaxa- tion period.

In the forced harmonic motion of industrial equipment, damping force lags by an angle $\pi/2$ compared with elastic force. For this reason, the maximum force F_{tr} that is transmitted to the foundation will be [2] and [3]:

$$F_{tr\max} = \sqrt{(kx_0)^2 + (c\omega x_0)^2} = x_0\sqrt{k^2 + c^2\omega^2} \quad (11)$$

4. MATHEMATICAL MODELING OF THE AIR COMPRESSOR UNIT

The mechanical model for vibration and isolation with elastic elements and damping viscous elements is pre- sented in Figure 3 [3] and [8].

The curves from the diagram of transmissibility coefficient (see Figure 3(b)) show the effect of damping forces, which is a reduction for the amplitude of forced vibration.

The vibration amplitude not becomes infinitely larger comparing with the case when $c = 0$ and when $\omega = \omega_n$. The other curves present maximum of amplitudes with decreasing values, if c increase, given by relationship:

$$\frac{\omega}{\omega_n} = \sqrt{1 - 2\left(\frac{c}{c_{cr}}\right)^2} \quad (12)$$

The machines for compressing air and its mounting scheme on the foundation are presented in Figure 4. A multi-cylinder compressor, as a five stage air compressor 1, is composed from two parts, law pressure stage (cy- linders I, II and III) and high pressure stage (cylinders IV and V), both being driven by an electric machine 2 via an elastic coupling S_f. All ensembles are fixed on a massive block of concrete 3 which is seating on the foun- dation 5. Between the block 3 and the foundation was interposed an elastic and damping structure 4.

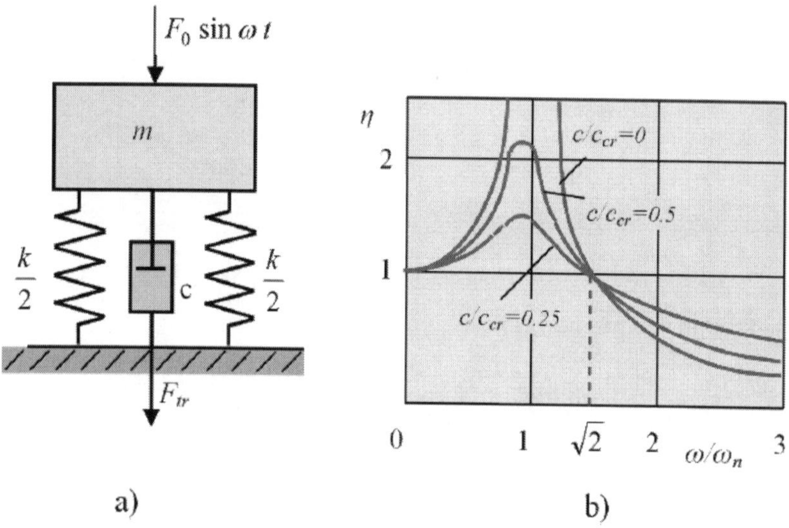

Figure 3. Experimental test plan, a) simplified mechanical model of the air compressor located on the foundation; b) diagram the coefficient of trans- missibility in function ratio pulsation.

The massive block foundation is used for two reasons: to machine axis line stiffening and to diminishing ma- chine-foundation vibration excited by residual inertial unbalanced forces $F_1, \cdots, F_5$, acting on machine [9]. The vibration levels depend on the dynamic characteristics of the elastic and damping structure 4 and on the in- ertial mass of the massive concrete block 3.

To determined the distribution of vibration was measured the vibration motions components, $w_{xi}(t)$, $w_{yi}(t)$ and $w_{zi}(t)$, in a lot of points P_i along the ensemble machine massive foundation. For the motion of a rigid body, the distribution of vibration occurs is [8] and [10].

$$\{w_i(t)\} = \{w_0(t)\} + [R] \cdot \{\varphi(t)\} \tag{13}$$

where $\{w_0(t)\}$ is the column vector matrix for the resultants vibration, and is given by:

$$\{w_i(t)\} = \{w_{xi}(t) \quad w_{yi}(t) \quad w_{zi}(t)\}^{T} \tag{14}$$

- $\{w_0(t)\}$ is the column vector matrix of vibration point P_i at radius r_i - compared to the geometric centre 0 of the structure;

- $\{\varphi(t)\}$ is the column vector matrix of "out of phase" at point P_i compared to the reference system X0YZ-chosen arbitrarily and is given by:

$$\{\varphi(t)\} = \{\varphi_x(t) \quad \varphi_y(t) \quad \varphi_z(t)\}^{T} \tag{15}$$

- $[R]$ is the matrix of position point P_i compared to the reference system X0YZ and is:

$$[R_i] = \begin{bmatrix} 0 & z_i & -y_i \\ -z_i & 0 & x_i \\ y_i & -x_i & 0 \end{bmatrix} \tag{16}$$

Distribution law for the rigid body vibration given by relationship (13), can be extended to the distribution of amplitudes:

$$\{W_i\}_{c,s} = \{W_0\}_{c,s} - [R] \cdot \{\Phi\}_{c,s} \tag{17}$$

where with the letters c and s are meaning the terms $\sin\omega t$ and $\cos\omega t$.

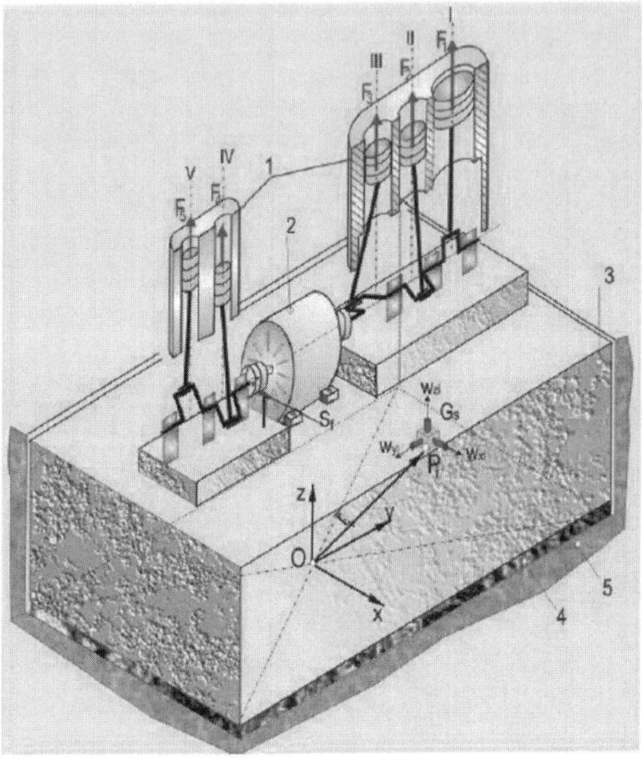

Figure 4. Mounting scheme of a machine for air compression (with more cylinders).

5. EXPERIMENTAL RESEARCHES

Experimental investigations have been focused on study of isolating vibration for two air compressor units from an oxygen station. Although the two aggregates were identical in terms of construction and were also placed on the identical foundations, one of them presented higher levels of vibration than the other.

Was emitted the hypothesis that one of aggregates had hidden defects at installation, which led to its prema- ture wear. But the two aggregates were new, and verification tests done before entering into service did not re- veal any hidden defect.

Has been also hypothesised that the concrete massive block interposed between aggregate and foundation, has "cracked" during the technological tests, and as result it was amplifying the disturbance forces generated by the mechanical (vibrating) system.

In the first phase was carried out a dimensional static verification of the two units, but no structural differ- ences were found between them. Tolerances of the active elements—pistons, cylinder shirts, rod-crank mechanism have been found to be into the prescribed limits. There were not found abnormal wear of these compo- nents and the mechanism rod-crank it is was working smoothly, evenly and without shocks at change of di- rection.

These results had supported the hypothesis that disruptive forces $F_1, \cdots, F_5$ are the same for both aggre- gates. To explain the large difference between the vibrations of the two units was placed the emphasis in the second phase of experimental research, it was performed integrity verification for the foundation. This was as- similated to a massive mass that must have the vibration motion of a rigid body. In the case of some hidden cracks, significant deviations from the rigid body anticipated movements will be present.

To determine the vibration distribution, we measured the components for the vibration movements $w_{xi}(t)$, $w_{yi}(t)$ and $w_{zi}(t)$—see relationship (14)—for several points P_i along the perimeter of aggregates placed on the solid concrete block of the foundation (see Figure 4).

The experimental measurements we made with a comprehensive installation type DELTA TRON®, have al- lowed the effective determination of vibration amplitudes, and out of phase according to the model shown in Figure 4. The vibration motion distribution for the two air compression machines is shown in Figure 5.

Were used 8 tri-axial transducers $T_{r1}, \cdots, T_{r8}$ for vibration monitoring, emplaced on the massive concrete block 3—in two parallel planes. It has been established one rectangular system X0YZ shown in the figure, as a reference basis for the reported measured vibration parameters.

The values for the vibration motion amplitude and out of phase for the first harmonic, corresponding to the two machines are given in the table below:

	Aggregate No. 1				Aggregate No. 2		
Parameter	$\cos(\omega t)$	$\sin(\omega t)$	MU	Parameter	$\cos(\omega t)$	$\sin(\omega t)$	MU
Wo_x	0.56	−0.84		Wo_x	1.77	−0.10	
Wo_z	−1.70	0.87	μm	Wo_z	−1.77	−0.68	μm
Wo_z	−8.35	2.88		Wo_z	−12.13	−3.03	
Φ_x	−4.40	−2.00		Φ_x	−6.44	−6.38	
Φ_y	−0.35	−0.13	μm/m	Φ_y	−0.52	−1.75	μm/m
Φ_z	0.04	−0.42		Φ_z	0.04	−0.42	

If we analyze these results, we cannot observe significant differences between the two aggregates. Therefore the hypotheses of premature wear, or the hypothesis of some cracks in the massive concrete block at one of the two units, were rejected.

Experimental researches performed in a longer time, showed however, random occurrence of some dangerous frequencies at different moments and speeds for the electric motor 2. Because the disturbance forces $F_1, \ldots, F_5$ are acting periodically on the foundation with the ωt pulsation, and because the force F_{tr} transmitted to the foundation did not exceed the maximum allowable limit, it was advanced the hypothesis of a variable damping force acting randomly in time. The coefficient of transmissibility determined by relation (5) has, therefore, a va- riable character.

This hypothesis was accepted by the oxygen station management, which has assumed responsibility for the foundation work quality. Due to the difficulties encountered during the work for gravel compaction, it was not followed exactly the technological recipe at one of the two foundations.

6. CONCLUSIONS

The study on vibration isolation has highlighted the importance of a strict compliance with the technical re- quirements for the foundation execution technology. To explain the large difference between vibration of the two aggregates, we focused on the integrity of the concrete foundation verify which as a massive mass must to has vibration of rigid body motions. In case of a possible hidden crack the significant deviation from the rigid body motions may occurs.

The damping forces variation during operation of the equipment, leads to some variations for the transmissibility coefficient values (see Figure 3(b)):

- For $\omega < \sqrt{2}\omega_n$, coefficient of transmissibility is higher than one, regardless of the ratio c/c_{cr}; which includes the case of resonance, viscous damping is advantageous because it reduces the transmissibility coefficient and therefore the amplitude of vibration;

- For $\omega > \sqrt{2}\omega_n$, coefficient of transmissibility is less than one, and it is lower for foundations with elastic elements only, and increases with the ratio c/c_{cr}; in this field, damping effect is unfavorable, so that for this situation—are preferred constructive solutions with materials with smaller elastic constant k.

- Whereas in the period starting and stopping, industrial equipments are passing through critical speeds, there is the possibility of the uncontrolled appearance of some dangerous vibrations; for this reason is recommended to use viscous damping with ratio c/c_{cr} as small as possible or at most equal to 0.5.

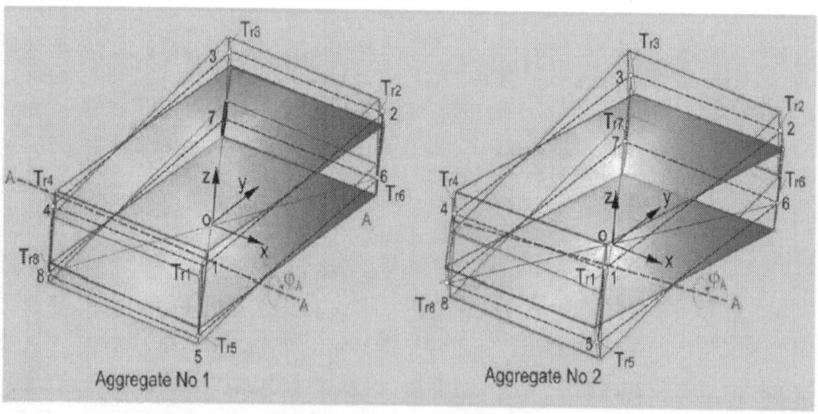

Figure 5. Scheme distribution of vibrational motion of the foundations for the two air com- pression aggregates.

REFERENCES

1. Achenbach, J.D. (1984) Wave Propagation in Elastic Solid. North Holland, New York.

2. Beards, C.F. (1995) Engineering Vibration Analysis with Application to Control Systems. Edward Arnold, London.

3. Blevins, R.D. (1979) Formulas for Natural Frequency and Mode Shape. Van Nostrand Reinhold, New York.

4. Cioara, T.Gh., Nicolae, I., Cires, I., Cristea, D., Cires, D. and Caplanescu, C. (2010) Consideration on Structure Vibra- tion Isolation Asymmetry. Two Funny Cases. Proceedings of the IMAC-XXVIII, Jacksonville, 1-4 February 2010.

5. Craig, R.R. and Kurdila, A. (2006) Fundamentals of Structural Dynamics. John Wiley and Sons, New York.

6. Silas, Gh. (1968) Mechanics. Mechanical Vibration. Didactic and Pedagogical Publishing, Bucharest.

7. Geradin, M. and Rixon, D. (1997) Theory and Application to Structural Dynamics. 2nd Edition, John Wiley and Sons, New York.

8. Hussey, M. (1983) Fundamentals of Mechanical Vibrations. Mac Milan Press Ltd., London.

9. Meirovitch, L. (2001) Fundamentals of Vibration. McGraw-Hill, New York.

10. Rades, M. (2008) Mechanical Vibration. Publishing Printech Bucharest.

CHAPTER 6

Intelligent Fault Detection of Retainer Clutch Mechanism of Tractor by ANFIS and Vibration Analysis

Ebrahim Ebrahimi[1], Payam Javadikia[2], Mohammad Hadi Jalili[1], Nasrolah Astan[1], Majid Haidari[3], Mojtaba Bavandpour[1]

[1] Department of Mechanical Engineering of Agricultural Machinery, Islamic Azad University, Kermanshah Branch, Kermanshah, Iran

[2] Department of Mechanical Engineering of Agricultural Machinery, Razi University, Kermanshah, Iran

[3] Department of Mechanical Engineering of Agricultural Machinery, Islamic Azad University, Takestan Branch, Takestan, Iran

ABSTRACT

In this study, ANFIS, as decision support system, is applied to detect the faults of MF 285 mechanism tractor clutch. Maintenance mechanisms include normal mode, rolling element failure, seal failure and attrition-based. Experiments were carried out in three speeds: 1000, 15,000, 2000 RPM and two conditions. The sensor was mounted vertically and horizontally. Vibrating spectrum of the time domain and the frequency of vibration data were obtained. Thirty-three statistical parameters of vibration signals in frequency domain and time were chosen as the sources attribute to detect errors. Finally, the top three features as input vectors to the ANFIS were evaluated. Using statistical parameters the performance of the system was calculated with the experimental data and training of ANFIS model. The system could not provide a seal to identify the fault. Regardless of the vibration data obtained from the classification of the seal, the overall classification accuracy of the ANFIS was 99.14% in the amount of 100% of the sensor installed vertically and horizontally. The results showed that this system can be used as an intelligent diagnosis system.

KEYWORDS

Fault Detection; Maintenance Clutch Mechanism; Vibration Analysis; Neuro-Fuzzy Inference Systems

1. INTRODUCTION

Rotating machines are the mainstay of several industrial applications worldwide. They can be found automobile, water industry, wind turbines, agricultural machinery, etc. Failures associated with gears represent the cause of extended outages and are typically caused by gradual deterioration and wear [1].

An important objective of condition-based maintenance (CBM) is to determine the optimal time for replacement or overhaul of a machine. The ability to accurately predict the remaining useful life of a machine system is crucial for its operation and can also be used to improve productivity [2].

Vibration analysis is a powerful tool in machine fault diagnosis. It is a way to check the status of a machine using a special tool. Troubleshooting machines can be more identified in the future and before the crash.

Fault diagnosis of mechanical or electronic components or systems is a subject of expert system applications. Today, neuro-fuzzy devices are emerged by combining network based on adaptive neural networks and fuzzy models. These adaptive networks, based on fuzzy inference system, are called ANFIS. In such systems the desired output will be achieved by means of human knowledge, in the form of if-then rules for the desired input. Dealing with systems that are not well-defined it is not true model the system based on conventional mathematical tools and component identification and therefore the neuro-fuzzy systems could be applied.

Many researchers [3-12] have employed various methuds to detect weak periodic signals using the sensitivity of system parameters. Some of these methods are shock pulse monitoring, crest factor analysis, kurtosis, spectrum analysis, envelope spectrum analysis, wavelet analysis, filtration of signal, etc.

Among the researches carried on machine condition monitoring with vibration analysis using fuzzy logic and neuro-fuzzy networks there are rotary machines fault diagnosis, troubleshooting automotive active suspension system and taking into account the non-linear behavior of hydraulic actuator and significantly the impaired controller, an engine control ventilation, fault diagnosis of bearing rotary, fault diagnosis of rolling bearing [4,7,12], tapered gearbox fault diagnosis [11].

In agricultural tractors, fractures in different parts of the power transmission due to the lack of early detection and location of damage and repair or replace the damaged part will be caused waste of time, cost more to repair and loose time for farming operations which often are duly carried. MF285 Tractor, is equipped with a two stage clutch. The two main types of clutches are necessary. The first resembles a driven plate clutch to power takeoff but the latter is based on the sheath power takeoff. The shaft makes the gear box to move while the pad moves through two gears toward the empowerment shaft. **Figure 1** illustrates the position of each component.

Failure on this part will lead to some fraction in other parts specially in gearbox of tractor, on condition of late diagnosis. The overall objective of this research is to present the intelligent detection of some parts of MF285 tractor clutch holder mechanism with high accuracy.

2. MATERIALS AND METHODS

Classical methods for fault diagnosis include measurable change of system, values of tolerance condition monitoring of machinery health state. This method requires the presence of a person qualified to comment on the health of the defective machine. This study intends to provide an intelligent fault diagnosis system, so that it could help to give a quick decision on machine safety to an expert. In this study the experiments on trial setup of MF285 tractor clutch holder mechanism in a single laboratory unit (**Figure 2**).

Intelligent system for Condition monitoring is generally, divided to three parts: data acquisition, signal processing and feature extraction and classification of defects. The proposed system for failure detection includes vibration signals collection in data acquisition, data transfer from time zone to frequency zone in signal processing, calculating the most important features in features extraction and finally data entrance to ANFIS and calculating the total classification accuracy.

2.1. Data Acquisition

Data was acquired in two accelerometer sensor direction and three rotation speeds include 1000, 1500 and 2000 RPM in safe mode on laboratory setup. The accelerometer was installed in two X and Y directions on the main holder. ARMA Software was used for this purpose and 10,240 voltage data were recorded and saved in an Excel file. Data collected for each sample took 10 s. Total 24 states and 130 vibration data for each state (780 for safe mode and 2340 for down mode and totally 3210 samples) were extracted.

Figure 1. The position of clutch holder mechanism (right) and internal components of clutch holder mechanism (left).

Different failures that may be happen on mechanism were occurred on different parts of clutch holder mechanism (**Figure 3**) and the vibration data acquired separately. **Table 1** shows the specifications of each failure.

Due to the breakdown in rotating machines, time domain signals vary in each fault conditions. Vibrational spectra taken during the time of each of the modes are obtained. A typical vibration spectrum obtained at a moment when the photograph is taken. **Figure 4** shows a sample of vibration spectrum obtained during data mapping taken momentarily.

Figure 2. Setup tools and parts used in the laboratory.

Figure 3. Clutch holder mechanisms failures: seal failure (top-right), roller failure (top-left), PTO shaft friction (bottom).

Table 1. Specifications of each failure.

Failure type	Failure profile
Seal failure	The inner seal is torn and the coil is damage
PTO shaft friction	Shaft diameter is reduced upto 10 percent
Roller failure	Three roller balls and the inner ring are destroyed

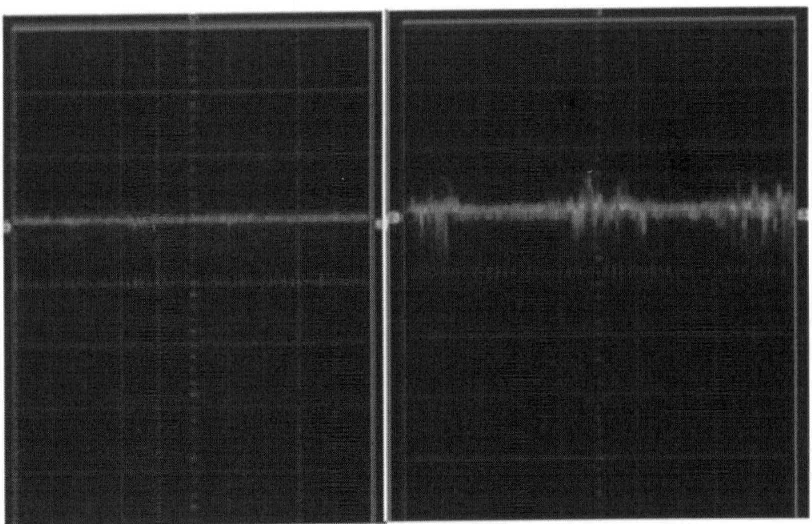

(a) (b)

Figure 4. Vibrationsignal from health condition (a) andvibrationsignal from fault condition (b) horizontal sensor and 1000 rmp.

2.2. Signal Processing

Signal processing and feature extraction are the most important parts of the monitoring process by which the system feature including safety or failure and the system type could be specified. One common way to provide useful features is to transfer signal from time domain to frequency domain. The analyzer, using the Fourier series, changes the analogous signals of time zone into digital data in frequency zone. For this purpose, at first program the necessary codes in the form of MATLAB. This Excel program contains vibration data and using the filter applied, the data was filtered the noises were eliminated and finally the time and frequency statistical parameters were calculated. So, some conversion feature of Fourier series (FFT) with 10240 points was calculated for each time signal. Also, the power spectrum density (PSD) and phase angle (FFT) of vibration signals were calculated using MATLAB.

Using MATLAB, the diagrams related to vibration spectrum in frequency domain in safe and failure modes were acquired to be analyzed and compared with each other (Figures 5 and 6).

2.3. Feature Extraction

The main purpose of feature extraction techniques is to present failure features in signals. In this research, the time domain data with 10240 formed the sample. Also, the frequency domain data points, which is PSD data and FFT phase angle were each 10,240. These signals were used to extract thirty-three features of safe and failure modes. Eleven features (T1 - T11) of time domain statistical parameters and rest of parameters (P1 - P11) and (A1 - A11) form the frequency domain statistical parameters.

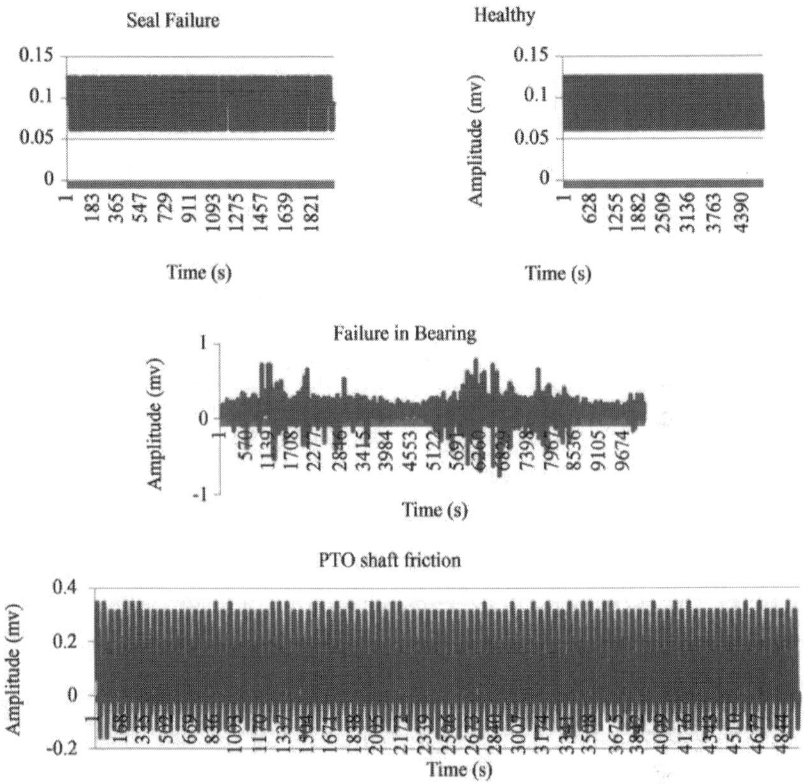

Figure 5. Typical vibration spectra obtained in time domain.

2.4. Fault Classification with ANFIS

For better definition of output result each of safe and fault modes have a code to classify the failures (**Table 2**).

Then in order to data model in ANFIS, 8 models (A - H) were considered for modeling in ANFIS and experimental data were placed in classification (**Table 3**).

During the initial data mapping and using seal fault in laboratory model, the vibration resulted from seal fault were felt less than safe mode and no significant difference was observed from initial comparison of data obtained from seal failure and safe mode and their vibration spectra. On the other hand, after investigating data related to seal fault, in order to select super features, less significant features from data on seal fault were observed. So, it was assumed that the present system could not have a good show of defects resulted from seal fault. Therefore, the data related to seal fault were eliminated in B and H models and other data were entered into the network.

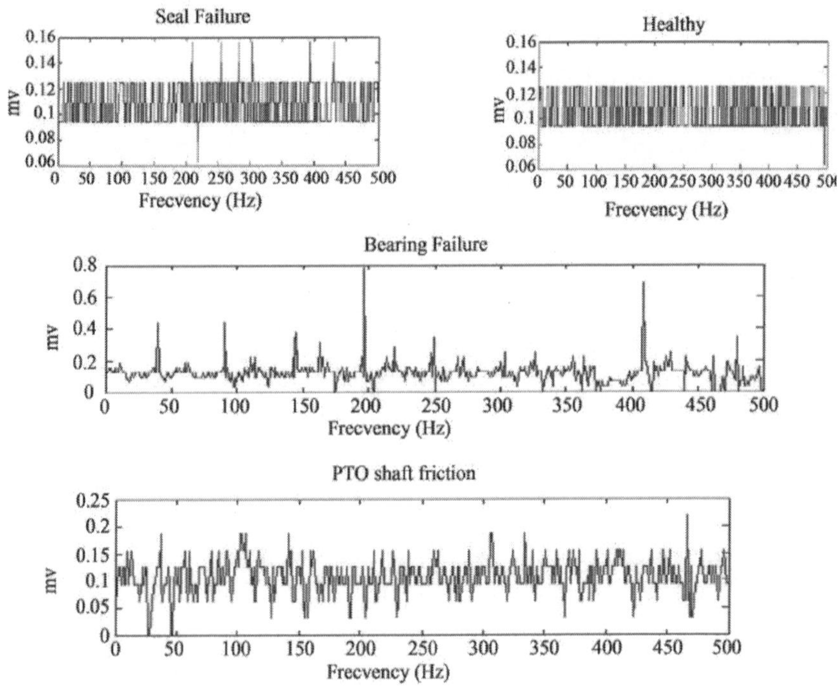

Figure 6. Frequency spectra obtained in frequency domain.

Table 2. Code definition for each safe and failure mode.

Mode	Safe	Roller fault	Seal fault	Axis friction
Defined code in classification	0	1	2	3

Table 3. Modes classification and input data to ANFIS.

Experiment model	Sensor status	(rpm)	Classification code	Number of experimental data	No. of training data	Input to ANFIS
A	Vertical	1000	3210	468	1092	A_9 P_{11} A_8
B	Vertical	1000	310	351	819	A_9 P_{11} A_8
C	Vertical	1500	3210	468	1092	A_9 P_{11} A_8
D	Vertical	2000	3210	468	1092	A_8 P_9 A_7
E	Horizontal	1000	3210	468	1092	T_5 A_9 A_8
F	Horizontal	1500	3210	468	1092	A_9 P_{11} A_8
G	Horizontal	2000	3210	468	1092	T_5 A_9 A_8
H	Horizontal	2000	310	351	819	T_5 A_9 A_8

3. RESULTS

The FFT diagram is the most applicative device for the vibration analysis. But much information on machine behavior may be eliminated in FFT diagram for various results like: some frequencies filtering, the adjacent frequencies overlap regarding the selective clarity. The neuro-adaptive techniques provide some method to make a phase modeling trend along with information learning from a data collection. In order to evaluate the proposed fault detection system the vibration data were placed in classification and 8 models were considered for modeling and entering the ANFIS network. The ANFIS feature vector was extracted for each model in different sensor modes and rotations per each three features. For the experiment models, in order to train and test the data, better significant features which made better results using the failure and try method were applied to enter the network. 70% data in each model were assigned to train and 30% were loaded to test in the network. To obtain better results the input and output membership features and different training periods were considered. Also, the initial value of 1% as the step size was formed for the parameters adaptation. After loading for all models, the failure diagram was drawn based on didactic periods. Then, the rules monitor form, fuzzy rules, fault detection topology and levels diagram were extracted for each separate model. Meanwhile, from the results obtained of ANFIS, the mean square error (MSE), sum of square error (SSE), correlation coefficient (R), mean absolute error (MAE), the degree of significance (P) and the prediction were obtained (**Table 4**).

By observing the result of ANFIS and counting correct answers, confusion matrix for the entire model was established. In **Table 5** the confusion matrix is illustrated for model B.

Diagonal elements in matrix represent the number of correct answers.

Overall classification accuracy, sensitivity and statistical are two statistical criteria to elevate the performance of the system in classification. These criteria are defined as follows:

Table 4. The best results of the ANFIS models in different experiment.

Model Testing	The results of the ANFIS			
	R	SSE	MAE	MSE
A	0.80	77.9	0.44	0.49
B	0.98	1.02	0.05	0.06
C	0.82	69.00	0.39	0.44
D	0.66	124.6	0.60	0.79
E	0.78	254.4	0.53	0.48
F	0.89	135.9	0.32	0.26
G	0.73	112.5	0.49	0.72
H	0.99	0.002	0.0015	2.44

Table 5. The confusion matrix for flaw detection in model B.

Output/Desired	Safe	Roller failure	Axis friction
Safe	39	0	0
Roller failure	1	38	0
Axis friction	0	0	39

Sensitivity equals to the number of correct positive decisions divided by total possible decisions. Total classification accuracy equals to the number of correct decisions divided by total possible decisions. Finally, the statistical criteria were obtained to detect system performance in classifying each model (**Table 6**).

Table 6. Statistical criteria and classification accuracy obtained in all test models.

Statistical criteria		Status	Model
Total percent	Sensitivity percent		
	54.8	0	
	100	1	
67.9	17.5	2	A
	100	3	
	100	0	
	97.43	1	
99.14	100	3	B
	43.6	0	
	100	1	
	43.6	2	
71.79	100	3	C
	30.76	0	
	100	1	
	64.1	2	
66.66	71.8	3	D
	15.38	0	
	82	1	
	0	2	
46.15	87.17	3	E
	46.15	0	
	100	1	
	33.33	2	
69.78	100	3	F
	51.28	0	
	100	1	
	30.76	2	
69.78	97.43	3	G
	100	0	
100	100	1	H
	100	3	

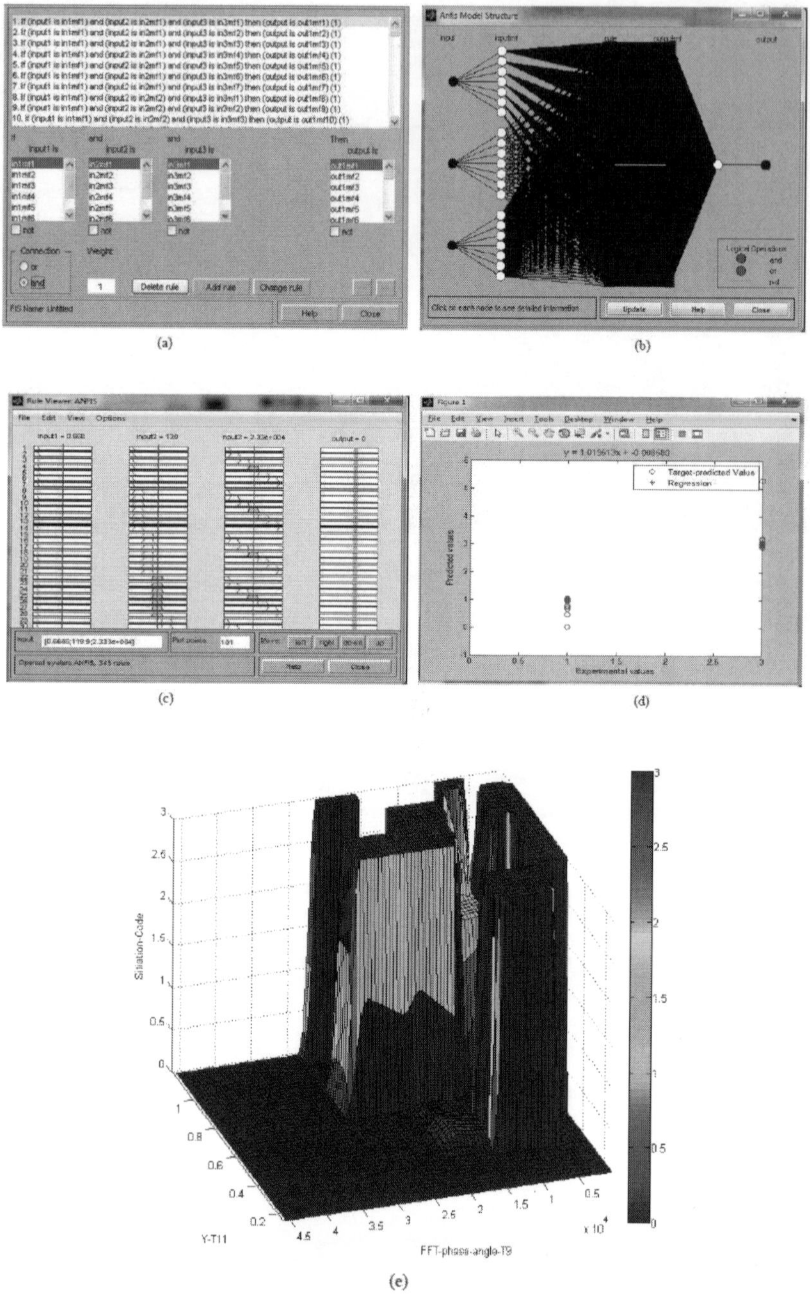

Figure 7. (a) The extraction of fuzzy rules; (b) Designed ANFIS topology; (c) Rules monitor, regression diagram on predictions; (d) The extraction of levels monitor in model B; (e) The topography of model.

In order to compare the success of failure detection of the models together, a description of overall classification accuracy in each case is shown as diagram (**Figure** 7). In model B the total classification accuracy were 99.14% and the highest amount of accuracy 100% was obtained in model D, which is desired. Also, the total classification mean accuracy of fault detection system with vertical sensor 76.41 and horizontal sensor 71.47 was obtained (**Figure 8**).

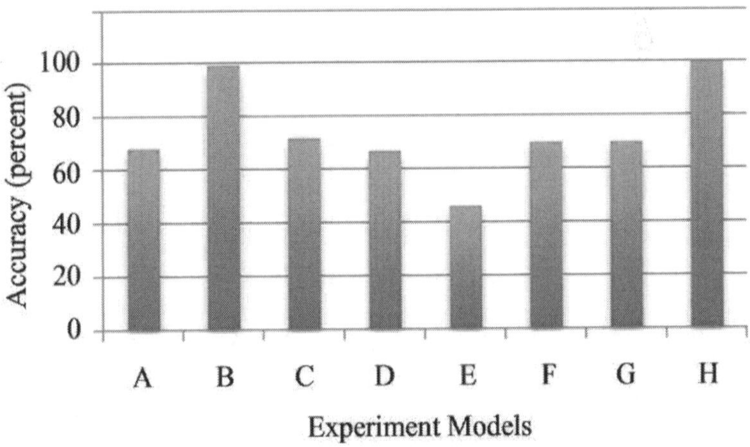

Figure 8. Diagram of total classification accuracy based on experiment models.

4. CONCLUSION

In this study, vibration data resulted from experimental setup were obtained using the data acquisition system. Statistical features of time and frequency domain were obtained for different failures detection and among them three super features were selected to enter the ANFIS network. 8 models were places in classification to detect the failure. In two models (models B and H), the seal fault was not placed in classification. After calculating statistical parameters, the total classification accuracy in model D and H were, respectively, 99.14

and100 percents. Comparing the results of total classification accuracy obtained from ANFIS models showed the assumption that the seal failure makes little vibration is true and the proposed fault detection system was not able to show the seal failure. As a whole, the results showed that using intelligent neuro-fuzzy logic could be an effective method to detect the failures in clutch holder mechanism.

ACKNOWLEDGEMENTS

The authors would like to thank Islamic Azad University, Kermanshah Branch for its support to this research.

REFERENCES

1. M. Elforjani, D. Mba, A. Muhammad and A. Sire, "Condition Monitoring of Worm Gears," Applied Acoustics, Vol. 73, No. 8, 2012, pp. 859-863.

2. H.-E. Kim, A. C. C. Tan, J. Mathew and B.-K. Choi, "Bearing Fault Prognosis Based on Health State Probability Estimation," Expert Systems with Applications, Vol. 39, No. 5, 2012, pp. 5200-5213.

3. D. G. Lewicki and H. J. Decker, "Transmission Bearing Damage Detection Using Decision Fusion Analysis," Glenn Research Center, Cleveland, Army Research Laboratory, NASA/TM, 2004, pp. 1-20.

4. G. Goddu, B. Li, M. Chow and J. Hung, "Motor Bearing Fault Diagnosis by Fundamental Frequency Amplitude Based Fuzzy Decision System," Development of Electrical and Computer Engineering, North Carolina University, 1998, pp. 1961-1965.

5. N. Hotwai, "Vibration Analysis of Faulty Beam Using Fuzzy Logic Techique," Department of Mechanical Engineering National Institute of Technology, Rourkela, 2009, pp. 23-27.

6. Ch. Kong, J. Ki, S. Oh and J. Kim, "Trend Monitoring of a Turbofan Engine for Long Endurance UAV Using Fuzzy Logic," KSAS International Journal, Vol. 9, 2008, pp. 64-70.

7. Z. Kiral and H. Karagülle, "Simulation and Analysis of Vibration Signals Generated by Rolling Element Bearing with Defects," Department of Mechanical Engineering, Dokuz Eylul University, Turkey, 2003, pp. 667-678.

8. Y. Lei, Z. Hea, Y. Zia and Q. Hua, "Fault Diagnosis of Rotating Machinery Based on Multiple ANFIS Combination with Gas," Department of Mechanical Engineering, Xi'an Jiaotong University, Xi'an, 2003, pp. 667-678.

9. Y. A. Lei, Z. He and Y. Zi, "A New Approach to Intelligent Fault Diagnosis of Rotating Machinery," Expert Systems with Applications, 2008, pp. 1593-1600.

10. R. Martins Marcal, K. Hatakeyama and A. Susin, "Managing Incipient Faults in Rotating Machines Based on Vibration Analysis and Fuzzy Logic," Electrical Engineering Department, 2006, pp. 1-6.

11. N. Saravanan and S. Cholairajan, "Ramachandran Vibration-Based Fault Diagnosis of Spur Bevel Gear Box Using Fuzzy Technique," Department of Mechanical Engineering, Amrita Vishwa Vidyapeetham, India, 2009, pp. 3119-3135.

12. J. H. Singonahalli and N. R. Iyer, "Etection of Roller Bearing Defect Using Expert System and Fuzzy Logic," Department of Mechanical Engineering, California University Sacramento, Sacramento, 1996, pp. 596-613.

CHAPTER 7

Vibration Control of a Plate Subjected to Impulsive Force by Plate-Type Dynamic Vibration Absorbers

Herve Sonfack Bouna, Blaise Romeo Nana Nbendjo*

Laboratory of Modelling and Simulation in Engineering,
Biomimetics and Prototype, Faculty of Sciences,
University of Yaounde I, Yaounde, Cameroon

ABSTRACT

This paper examines modelling of the dynamics of a plate by plate type dynamics vibration absorber subjected to a localized periodic impulsive excitation. An analytical solution of the modal equation is proposed and validated using direct numerical simulation of the basic equations. The basics equations are solve numerically using fourth order Runge Kutta algorithm. Various types of dynamic absorbing plate are tested to optimize the control efficiency. Particular attentions have been paid on the effects of localization of external forces on the dynamics response of the system under control. Ours findings demonstrate that a good achievement of control strategy should follow the above mentioned analysis.

KEYWORDS

Plate; Impulsive Force; Vibration Control

1. INTRODUCTION

The problems of reduction of structural vibration induced by external force have constituted a great source of interest in recent years [1-4]. These excitations can be natural such as earthquakes, wind etc. or due to mechanical action. In both cases, this can lead to premature destruction of the system. One of the most famous examples illustrating these remarks is found in the Tacoma Narrows Bridge (USA, 1940) which was destroyed following an excitation by the wind. Several solutions have then been considered for this purpose. They range from simple archaic solutions such as destruction of these structures, to more modern solutions using techniques of vibration control which offer the advantage of being both less costly and more effective. Therefore, investigating vibration control of a plates become important in structures and mechanical systems design. To cope with these disturbances, many vibration control techniques have been developed such as sandwich control [5-8], electro-mechanical control [9-11], piezoelectric control [12,13], magneto rheological control [14], opto-electromechanical control [15,16].

In this paper, we are interested in the enhancement of viscoelastic control for a rectangular plate subjected to periodic impulsive excitation. The device is a three layers viscoelastic core subjected to localized periodic impulsive excitation. It is almost two rectangular plates of the same surfaces, hinged at both ends and possibly be made of different materials. The system is composed of two plates: One so-called main plate is subjected to impulsive external excitations located at point of contact, is coupled to another said dynamic absorbing plate by a viscoelastic coupling system consisting of identical springs and dampers with spring constant k and damping coefficient c respectively as shown in **Figure 1**. Therefore, several configurations of the coupling system can be used. We consider here the case of uniform distribution viscoelastic coupling.

In Ref. [17], Zhang et al. proposed to stabilize an elastic plate using viscoelastic boundary condition as feedback control. Hongpan et al. [18] investigated vibrations control of plate through electro-magnetic constrained layer damping which consists of, electromagnet layer, permanent layer and viscoelastic damping layer. Moreover, Aida et al. [6,7] have modelled the dynamic and optimized control by sandwich of a plate (and beam) under sinusoidal periodic excitation. They have shown that the effectiveness of this control is more effective as the number of springs and shock absorbers becomes higher. However, model using periodic sinusoidal excitation does not relate enough to processes occurring in real time. Therefore, as a first approach to overcome this issue localized periodic impulsive excitations are considered. Additionally, we identify and define the characteristics of the dynamic absorbing plate over the dynamic responses of the main plate.

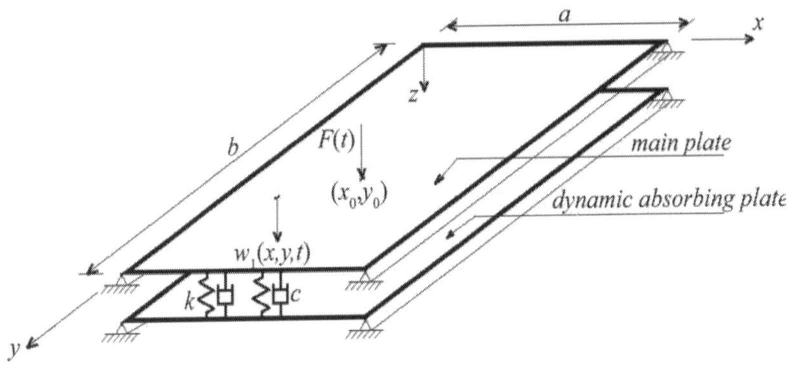

Figure 1. Main plate and dynamic absorbing plate with uniformly distributed connecting spring and dampers.

The second part of this paper deals with the derivation of the equations of motion of the main plate with hinged ends subjected to localized impulsive excitations. In the third part, analytical amplitudes of vibration are determined. Also, comparison between analytical and numerical amplitudes of vibration is provided. The fourth part is reserved to various configurations leading to the enhancement of the control process. We conclude our work in the last section.

2. MATHEMATICAL MODELLING OF THE SYSTEM

2.1. Modelling of Localized Periodic Impulsive Excitation

This type of load on a plate corresponds for example to the actions of generators or some robots. Both machines can be regarded as a mass device M, located on the plate by its coordinates (x_0, y_0). These devices in addition to their own weight per unit area $f_{\text{weight}}(x,y,t)$, impose at the point (x_0, y_0) of the plate, at regular time intervals equal to T, a surface force $f_{\text{impact}}(x,y,t)$ whose intensity (I) may change over time. So, the external excitation $F(x,y,t)$ is a combination of these two efforts:

$$F(x,y,t) = f_{\text{weight}}(x,y,t) + f_{\text{impact}}(x,y,t) \tag{1}$$

where,

$$f_{\text{weight}}(x,y,t) = P\delta(x-x_0)\delta(y-y_0) \tag{2}$$

$P = \frac{Mg}{S}$ while $f_{\text{impact}}(x,y,t)$ is defined literally by several models.

In general, the impact force is modelled as a rectangular pulse or as a Gaussian function whose integral over the duration of the impact is equal to the impulse of impact. The duration of the impact depends on the nature of contact between the exciter and the plate, the material they are made off and their dimensions. Therefore, minimizing the impact force is equivalent to minimizing the impulse of impact [19]. The case of rectangular plate is address in this study.

This model of excitation assume a force occurring at regular intervals of time equal to T and can be described as follows:

$$f_{\text{impact}}(x,y,t) = \begin{cases} 0 & \text{si} \quad t \in [(i-1)T + \Delta t, iT[\\ I\delta(x-x_0)\delta(y-y_0) & \text{si} \quad t \in [iT, iT + \Delta t[\end{cases} \tag{3}$$

with $i = 1,2,\cdots,N$.

N stand for the total number of impact while Δt represents the duration of the pulse.

The presence of the Dirac delta functions $\delta(x-x_0)$ and $\delta(y-y_0)$ denote the local character of the impulsive force. If the pulse duration Δt becomes too small as to be neglected, this model can be replaced by the model of the Dirac delta function giving by :

$$f_{\text{impact}}(x,y,t) = \left[\sum_{i=1}^{N} I\delta(t-iT)\right]\delta(x-x_0)\delta(y-y_0) \quad \forall t \in \mathbf{R} \tag{4}$$

2.2. Equations of Motion of the Plate under Control

The main plate (see **Figure 1**) which is under the same boundary conditions with the dynamic absorbing plate is attached to this one by springs and dampers distributed uniformly (see Ref. [6]).

The equations of motion of flexural vibration for the main and dynamic absorbing plates with uniform flexural rigidity and uniform mass per unit area are expressed as

$$\begin{cases} m_1 \frac{\partial^2 w_1}{\partial t^2} + D_1 \left[\frac{\partial^4 w_1}{\partial x^4} + 2\frac{\partial^4 w_1}{\partial x^2 \partial y^2} + \frac{\partial^4 w_1}{\partial y^4} \right] \\ +\delta_1 \frac{\partial w_1}{\partial t} + c\left(\frac{\partial w_1}{\partial t} - \frac{\partial w_2}{\partial t} \right) + k(w_1 - w_2) \\ = \left[P + \sum_{i=1}^{N} I\delta(t - iT) \right] \delta(x - x_0)\delta(y - y_0) \\ m_2 \frac{\partial^2 w_2}{\partial t^2} + D_2 \left[\frac{\partial^4 w_2}{\partial x^4} + 2\frac{\partial^4 w_2}{\partial x^2 \partial y^2} + \frac{\partial^4 w_2}{\partial y^4} \right] \\ +\delta_2 \frac{\partial w_2}{\partial t} + c\left(\frac{\partial w_2}{\partial t} - \frac{\partial w_1}{\partial t} \right) + k(w_2 - w_1) = 0 \end{cases} \quad (5)$$

where the deflections of the main and dynamic absorbing plates are denoted by $w_1(x,y,t)$ and $w_2(x,y,t)$ respectively. m_1 and m_2 are their mass per unit area, δ_1 and δ_2 their damping coefficients, D_1 and D_2 their flexural rigidity, k and c are the spring constant and the damping coefficient of connecting springs and dampers between the main and dynamic absorbing plates, respectively. I is the intensity of the periodic impulsive load, T the exciting period, P the weight per unit area of the device creating the impulsive load, δ is the Dirac delta function and (x_0, y_0) coordinates of the load position.

Let's denote ρ_1 and ρ_2 the densities, h_1 and h_2 the thicknesses, E_1 and E_2 the Young's modulus, ν_1 and ν_2 the Poisson's ratios of the main and dynamic absorbing plates respectively. It comes whereas:

$m_1 = \rho_1 h_1$, $m_2 = \rho_2 h_2$, $D_1 = \frac{E_1 h_1^3}{12(1-\nu_1^2)}$, and

$$D_2 = \frac{E_2 h_2^3}{12(1-\nu_2^2)}$$

While taking into account the boundaries conditions, the following approximate functions can be used for the solutions of equations of the system (5):

$$w_1(x,y,t)=\sum_{n=1}^{\infty}\sum_{m=1}^{\infty}W_{nm}(x,y)\chi_{1nm}(t) \tag{6}$$

$$w_2(x,y,t)=\sum_{n=1}^{\infty}\sum_{m=1}^{\infty}W_{nm}(x,y)\chi_{2nm}(t) \tag{7}$$

in which $\chi_{1nm}(t)$ and $\chi_{2nm}(t)$ are unknown functions of the time of the main and dynamic absorbing plates respectively and $W_{nm}(x,y)$ the normalized eigenfunction of mode (n,m) of the both plates given by :

$$W_{nm}(x,y)=\sin\left(\frac{n\pi x}{a}\right)\sin\left(\frac{m\pi y}{b}\right) \tag{8}$$

(n,m) is a natural mode with n and m nodal lines lying the xand y-directions, respectively, including the boundary as the nodal lines. Substituting Equations (7) and (8) into Equation (5), according to the orthogonality of the eigenfunction gives

$$\begin{cases} m_1\ddot{\chi}_{1nm}+m_1\omega_{1nm}^2\chi_{1nm}+\delta_1\dot{\chi}_{1nm} \\ +c\left(\dot{\chi}_{1nm}-\dot{\chi}_{2nm}\right)+k\left(\chi_{1nm}-\chi_{2nm}\right) \\ =\dfrac{4}{ab}W_{nm}(x_0,y_0)\left[P+\sum_{i=1}^{N}I\delta(t-iT)\right] \\ m_2\ddot{\chi}_{2nm}+m_2\omega_{2nm}^2\chi_{2nm}+\delta_2\dot{\chi}_{2nm} \\ +c\left(\dot{\chi}_{2nm}-\dot{\chi}_{1nm}\right)+k\left(\chi_{2nm}-\chi_{1nm}\right)=0 \end{cases} \tag{9}$$

ω_{1nm} and ω_{2nm} are natural frequencies of modes (n,m) of the main and dynamic absorbing plates respectively and are derive as follows:

$$\omega_{1nm}=\sqrt{\frac{D_1}{m_1}}\left(\left(\frac{n\pi}{a}\right)^2+\left(\frac{m\pi}{b}\right)^2\right) \tag{10}$$

$$\omega_{2nm} = \sqrt{\frac{D_2}{m_2}\left(\left(\frac{n\pi}{a}\right)^2 + \left(\frac{m\pi}{b}\right)^2\right)} \tag{11}$$

In non dimensional form, Equation (5) becomes,

$$\begin{cases} \ddot{z}_{1nm} + z_{1nm} + \beta_{1nm}\dot{z}_{1nm} + C\left(\dot{z}_{1nm} - \dot{z}_{2nm}\right) \\ +K\left(z_{1nm} - z_{2nm}\right) = P_0 + \sum_{i=1}^{N} I_0 \delta\left(\tau - iT_0\right) \\ \ddot{z}_{2nm} + \gamma z_{2nm} + \beta_{2nm}\dot{z}_{2nm} + \alpha C\left(\dot{z}_{2nm} - \dot{z}_{1nm}\right) \\ +\alpha K\left(z_{2nm} - z_{1nm}\right) = 0 \end{cases} \tag{12}$$

where

$$z_{1nm} = \frac{\chi_{1nm}}{l},\ z_{2nm} = \frac{\chi_{2nm}}{l};$$

$$\tau = t\omega_{1nm},\ T_0 = T\omega_{1nm};\ \alpha = \frac{m_1}{m_2},$$

$$\gamma = \left(\frac{\omega_{2nm}}{\omega_{1nm}}\right)^2 = \frac{m_1}{m_2}\frac{D_1}{D_2} = \alpha\frac{D_1}{D_2};$$

$$\beta_{1nm} = \frac{\delta_1}{m_1\omega_{1nm}},\ \beta_{2nm} = \frac{\delta_2}{m_2\omega_{1nm}};\ C = \frac{c}{m_1\omega_{1nm}},$$

$$K = \frac{k}{m_1\omega_{1nm}^2};\ P_0 = \frac{4}{m_1 abl\omega_{1nm}^2} PW_{nm}\left(x_0, y_0\right),$$

$$I_0 = \frac{4}{m_1 abl\omega_{1nm}^2} IW_{nm}\left(x_0, y_0\right)$$

where l a reference length.

3. ANALYTICAL SOLUTION OF THE MODAL EQUATIONS

Sine the most path of the energy of the system is embedded in the first mode of vibration, from Equation (9) we then expressed

$$\begin{cases} \ddot{z}_1+\beta\dot{z}_1+z_1+C\left(\dot{z}_1-\dot{z}_2\right)+K\left(z_1-z_2\right) \\ =P_0+\sum_{i=1}^{N}I_0\delta(\tau-iT_0), \\ \ddot{z}_2+\beta\dot{z}_2+z_2+C\left(\dot{z}_2-\dot{z}_1\right)+K\left(z_2-z_1\right)=0. \end{cases} \quad (13)$$

where

$m=m_1=m_2$; $D=D_1=D_2$;

$\omega=\omega_{1,11}=\omega_{2,11}=\sqrt{\frac{D}{m}\left(\left(\frac{\pi}{a}\right)^2+\left(\frac{\pi}{b}\right)^2\right)}$;

$\beta=\beta_{1,11}=\beta_{2,11}=\frac{\delta}{m\omega}$; $\alpha=1$ and $\gamma=1$; $C=\frac{c}{m\omega}$

and $K=\frac{k}{m\omega^2}$;

$P_0=\frac{4}{ablm\omega^2}PW_{11}\left(x_0,y_0\right)$ and

$I_0=\frac{4}{ablm\omega^2}IW_{11}\left(x_0,y_0\right)$; $Z_1=Z_{1,11}$ and $Z_2=Z_{2,11}$.

The resolution of Equation (13) is not straightforward. Hence to overcome such a situation, it useful to assume new variables as following:

$$\eta_1 = z_1 + z_2 \tag{14}$$

$$\eta_2 = z_1 - z_2 \tag{15}$$

Adding the two set of Equations (13) gives the following

$$\ddot{\eta}_1 + \beta\dot{\eta}_1 + \eta_1 = P_0 + I_0 \sum_{i=1}^{N} \delta(\tau - iT_0) \tag{16}$$

Subtracting the two set of Equations (13) yields to the following :

$$\ddot{\eta}_2 + (\beta + 2C)\dot{\eta}_2 + (1 + 2K)\eta_2 = P_0 + I_0 \sum_{i=1}^{N} \delta(\tau - iT_0) \tag{17}$$

The resolution of the two differential Equations (16) and (17) permits to find the solutions of (13) through the following expressions

$$z_1 = \frac{1}{2}(\eta_1 + \eta_2) \tag{18}$$

$$z_2 = \frac{1}{2}(\eta_1 - \eta_2). \tag{19}$$

Thus, Equations (16) and (17) can be rewritten as follows

$$\ddot{\eta}_j + 2\lambda_j\dot{\eta}_j + \omega_j^2\eta_j = f(\tau) \quad j = 1,2, \tag{20}$$

where

$$f(\tau) = P_0 + I_0 \sum_{i=1}^{N} \delta(\tau - iT_0)$$

$2\lambda_1 = \beta$, $\omega_1^2 = 1$; $2\lambda_2 = \beta + 2C$, $\omega_2^2 = 1 + 2K$

The solutions of Equation (20) are then derived and given by

$$\eta_j(\tau)=\left[A_j\cos\Omega_j\tau+B_j\sin\Omega_j\tau\right]\exp\left(-\lambda_j\tau\right)+\eta_{jP},$$
$$j=1,2 \tag{21}$$

in which A_j and B_j represent the constants of integration, η_{jP} are the particular solutions of Equation (20), and Ω_j are the vibration frequencies expressed as

$$\Omega_j^2=\omega_j^2-\lambda_j^2 \tag{22}$$

The shape of the external excitation $f(\tau)$ suggests a treatment at intervals in oder to determine the particular solutions η_{jP} [20]. So, while considering the realistic case of excitation for which the pulse is not actually instantaneous as advocated by the Dirac delta function, it seems worthy to consider the three following main in order to well characterize the dynamics of the plates.

- Phases I: $\forall\tau\in\left[(i-1)T_0+\Delta\tau;iT_0\right[$

(The dynamics responses are considered between the impacts of order $i-1$ and i)

Here, $f(\tau)=P_0$; therefore :

$$\eta_{jP}=\frac{P_0}{\omega_j^2} \tag{23}$$

- Phases II: $\tau\in\left[iT_0;iT_0+\Delta\tau\right[$

(The dynamics response are considered during the impacts of order i)

Here, one rather has $f(\tau)=P_0+I_0$; so :

$$\eta_{jP} = \frac{P_0 + I_0}{\omega_j^2}. \tag{24}$$

- Phase III: $\forall \tau \in [NT_0 + \Delta\tau; +\infty[$

(The dynamics response are considered after all N impacts)

This phase of the process can be discerned as an element of the first phases for $i = N+1$, corresponding to the dynamics of the plates between the impact of order N and the frictional impact of order $N+1$. The particular solutions of this phase are given by Equation (23). Therefore, the solutions z_1 and z_2 are finally obtained and given as

$$\begin{aligned} z_1(\tau) &= [A_1 \cos\Omega_1\tau + B_1 \sin\Omega_1\tau]\exp(-\lambda_1\tau) \\ &+ [A_2 \cos\Omega_2\tau + B_2 \sin\Omega_2\tau]\exp(-\lambda_2\tau) \\ &+ \frac{1}{2}(\eta_{1P} + \eta_{2P}) \end{aligned} \tag{25}$$

$$\begin{aligned} z_2(\tau) &= [A_1 \cos\Omega_1\tau + B_1 \sin\Omega_1\tau]\exp(-\lambda_1\tau) \\ &- [A_2 \cos\Omega_2\tau + B_2 \sin\Omega_2\tau]\exp(-\lambda_2\tau) \\ &+ \frac{1}{2}(\eta_{1P} - \eta_{2P}) \end{aligned} \tag{26}$$

The particular solutions η_{1P} and η_{2P} are given by expressions (21) according to the state of vibration phase. The constants of integration A_1, A_2, B_1 and B_2 are determined by the initial conditions of a current phase of the movement that are different from the final condition of the previous phase. Thus, the knowledge of the initial conditions of the first phase of the movement, that is also the initial conditions of the problem, permits to get closer to those of the other phases and therefore to determine the solutions of the problem at all times.

In the uncontrolled case ($K = C = 0$), it comes the following:

$$\lambda_1 = \lambda_1 = \lambda = \frac{\beta}{2},$$

$$\omega_1^2 = \omega_2^2 = \omega^2 = 1,$$

$$\Omega_1 = \Omega_2 = \Omega = \sqrt{\omega^2 - \lambda^2} = \sqrt{1 - \lambda^2},$$

$$\eta_{1P} = \eta_{2P} = \eta_P.$$

The two plates then become uncoupled and the solutions are now given by

$$z_{1nc}(\tau) = \left[A' \cos\Omega\tau + B' \sin\Omega\tau\right]\exp(-\lambda\tau) + \eta_P, \tag{27}$$

$$z_{2nc}(\tau) = \left[C' \cos\Omega\tau + D' \sin\Omega\tau\right]\exp(-\lambda\tau), \tag{28}$$

in which A', B', C' and D' are the new constants of integration.

The control is efficient when the maximum value of z_1 is lower that the one of z_{1nc}. This criteria enable to foresee the effectiveness of the control strategy.

4. NUMERICAL ANALYSIS

4.1. Validation of the Analytical Results

We compare graphically, the analytical and numerical solutions for several values of parameters of the problem. As an example, we consider the main and dynamic absorbing plates identical made of steel with the following parameters :

Density of material $\rho = 7850 \text{ Kg/m}^3$,

Young's modulus $E = 2.1\times10^{11}\ \mathrm{N/m^2}$ Poisson's ratios $\nu = 0.3$ Damping coefficient $\delta = 85.75\ \mathrm{N\cdot s/m}$ Width a = 0.9 mLength b = 1.5 mThickness h = 0.002 m.

These quantities permit to find the dimensionless parameter $\beta = 0.105$ and the free frequencies of the two plates as $\omega = 51.87\ \mathrm{Hz}$.

The fourth order Runge Kutta algorithm is used to computed the numerical solutions. Figures 2 and 3 display the time histories of the maximum amplitude of vibration in other to validate the result of analytical investigation. These figures show an excellent agreement between numerical and analytical results. The slight shifts observed for each impact are probably caused by sudden numerical fluctuations due to the discontinuity of the external excitation.

4.2. Enhancement of Control Process

Finding accurate condition that allows optimization of the control strategy is of interest. Thus we have at one's disposal, several possibilities to choose the type of absorbing plate and among which the following:

- Steel plate: $\nu_2 = 0.3$ $E_2 = 2.1\times10^{11}\ \mathrm{N/m^2}$, $\rho_2 = 7850\ \mathrm{Kg/m^3}$
- Tungsten plate: $\nu_2 = 0.3$ $\rho_2 = 7850\ \mathrm{Kg/m^3}$, $\rho_2 = 19300$ Kg/m^3
- Aluminium Plate (7075): $E_2 = 0.7\times10^{11}\mathrm{N/m^2}$, $\nu_2 = 0.35$, $\rho_2 = 2810\ \mathrm{Kg/m^3}$
- Concrete plate (E20): $E_2 = 0.2\times10^{11}\ \mathrm{N/m^2}$, $\nu_2 = 0.15$, $\rho_2 = 2400\ \mathrm{Kg/m^3}$
- Wood plate (plywood): $E_2 = 0.055\times10^{11}\ \mathrm{N/m^2}$, $\nu_2 = 0.2$, $\rho_2 = 800\ \mathrm{Kg/m^3}$

Each of these plates has thickness h_2 and their coefficients of dissipation δ_2, chosen such that the dimensionless parameter β_2 is always as $\beta_2 = \beta_1 = 0.105$. The initial conditions and the external force of excitation are the same that in the section (4.1); but with $I_0 = 4.$ and $P_0 = 0.$

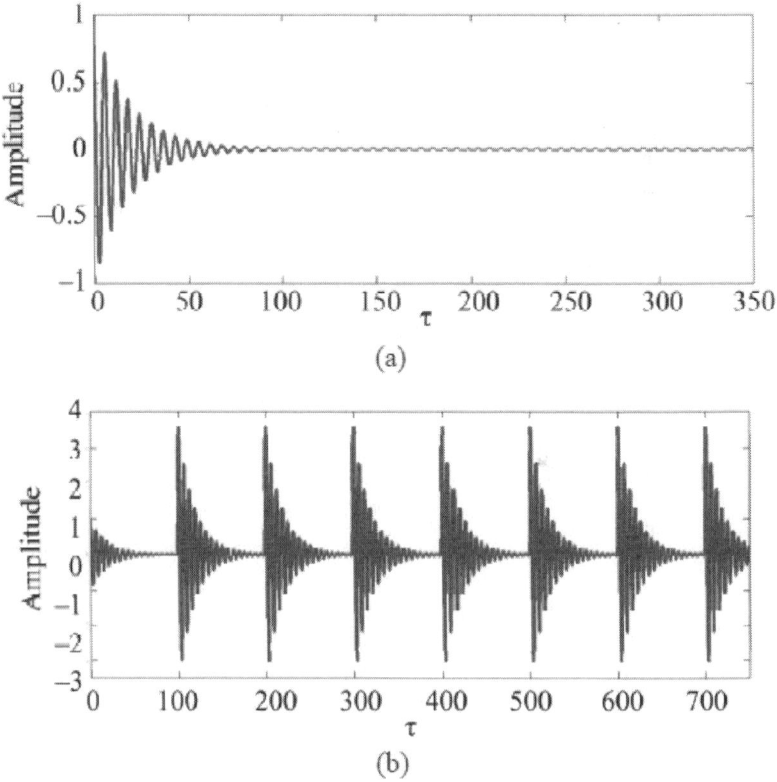

Figure 2. Times histories of maximum amplitudes of vibrations: Comparison between the analytical (red line) and numerical (discontinuous blue line) solutions; (a) $I_0 = 0$, $P_0 = 0$; (b) $I_0 = 4$, $P_0 = 0$.

4.2.1. Control Efficiency

The dynamic absorbing plate here is made of steel whose thickness is $h_2 = h_1 = 0.002$ m. The evolution of amplitude plotted in **Figure 4** shows that the amplitude of vibration under control is lower than the one without control. Then control is indeed efficient for the control parameters chosen. The possibility of the efficiency of the control being already pointed out for particular values of parameters of control. For that aim, we investigate separately their effects by considering other type plate having different physical characteristics compared to the one studied above.

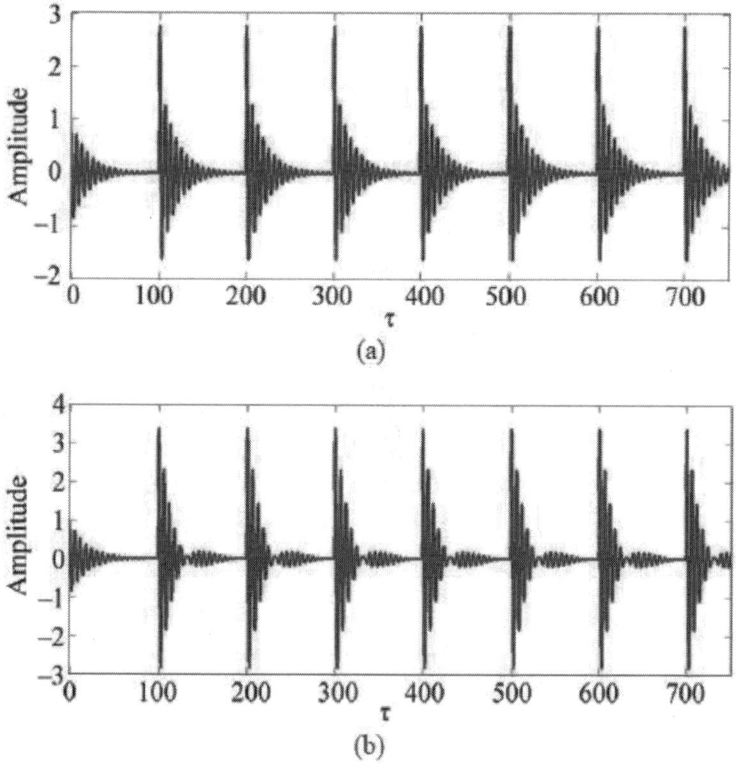

Figure 3. Times histories of the maximum amplitude of vibration: Comparison between the analytical (red line) and numerical (discontinuous blue line) solutions for $I_0 = 4$ and $P_0 = 0$; (a) $K = 0$, $C = 0.5$; (b) $K = 0.1$, $C = 0$.

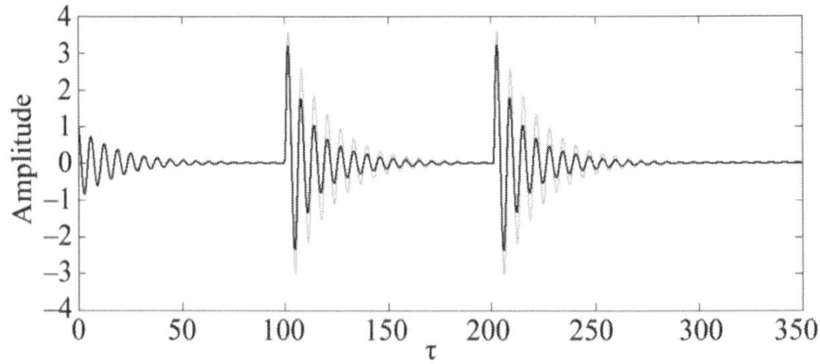

Figure 4. Times histories of the maximum amplitude of vibration: Comparison between the amplitude responses of the main plate in uncontrolled case (gray line) and in controlled case (black line) for K = 0.1, C = 0.1.

4.2.2. Effect of the Control Parameters

Here, we analyze how much the nature of absorbing plate impacts the control process. All the absorbing plates used here are taken such that the thickness $h_2 = 0.002$.

1) Effect of viscous coupling

Figure 5 shows that the control is efficient for all values of C since the amplitude of the control system is less than the one of the uncontrolled system (C = 0). This effect is particularly noticeable for higher values of C. In addition, this damping appears to depend on the material of the dynamic absorbing plate, and increases with the Young's modulus.

2) Effect of elastic coupling

As shown for the viscous control, the elastic control (see **Figure 6" target="_self"> Figure 6**) is also efficient for all values of K. Moreover, this efficiency increases as K becomes higher. In general, the absorbing plate made in steel gives more better protections compared to those made wood, concrete and aluminium.

3) Effect of viscoelastic coupling

A generalization shown in **Figure 7** reinforce our previous analysis revealing that the amplitudes of the vibrations decrease when the coupling becomes more and more robust. In other words, this happen when the coefficient of dissipation of the shock absorbers and the constant of stiffness of the springs increase.

4.2.3. Effects of the Characteristics of the Dynamic Absorbing Plate

In this subsection, the influences of the thickness and the Young's modulus of the dynamic absorbing plate on the amplitude of vibrations are investigated.

From **Figure 8** it appears that the vibration amplitudes decrease with the thickness h_2 and Young's modulus E_2 of the dynamic absorbing plate and stretches toward an unique limit when h_2 increases considerably. Furthermore, the curves exhibit peaks for certain values of the ratio h_2/h_1. The analysis of the position of this peak for the dynamic absorbing plate made in steel for which $\frac{h_2}{h_1}=1$, predicts that such behavior is observed when the vibration frequencies of the two plates are equal. To test this hypothesis as well as the effect of Young's modulus of dynamic absorbing plate, we plot the evolution of the amplitude of vibration of the main plate according to the vibration frequency of the dynamic absorbing plate for several values of its mass per unit area m_2 (see **Figure 9**).

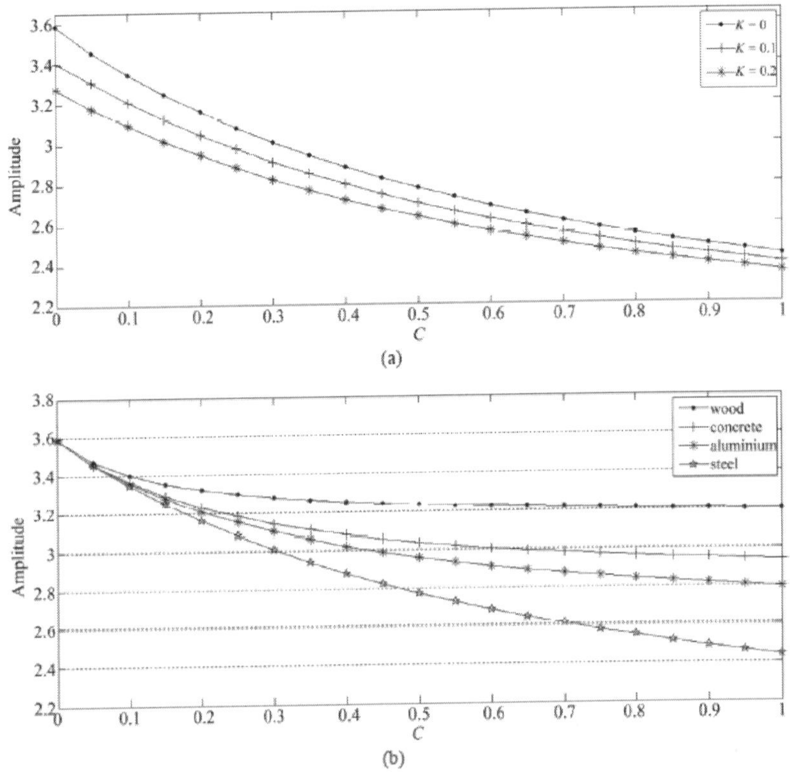

Figure 5. Effects of the viscous coupling on the amplitude response of the system: (a) For absorbing plate made of steel; (b) For absorbing plate made of wood, concrete, aluminium and steel.

The peak of amplitude occurs when the vibration frequencies are equal. In addition, the amplitude of vibration decreases with the vibration frequency ω_2 of the dynamic absorbing plate and with E_2 for which it has a linear dependency.

4.2.4. Effect of the Position of the Excitation

One interesting situation is the influence of the position of excitation on the main plate. For that aim, it has been plotted as shown in **Figure 10**, the three dimensional graph representing the maximum amplitude of vibration as function of coordinates position (x_0, y_0) of external force. It appears generally with and without control that, the amplitudes of vibration gradually decrease when the excitation device goes from the center of the plate to the edges. As a consequence, one way to reduce the amplitude of vibration could be the choice of localize position of the excitation.

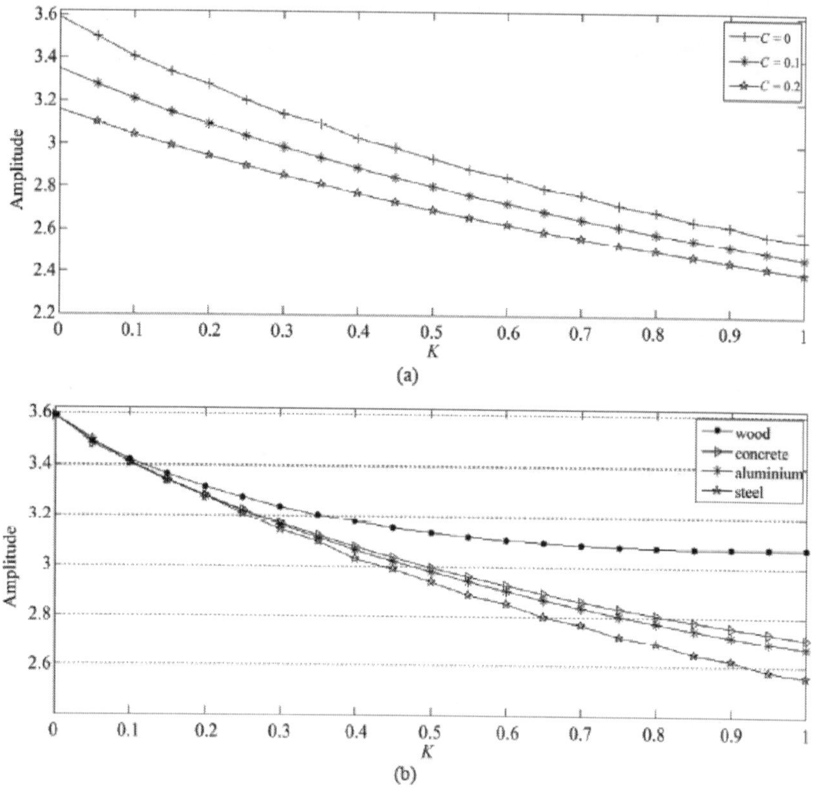

Figure 6. Effect of the elastic coupling on the response of the system: (a) For absorbing plate made of steel; (b) For absorbing plate made of wood, concrete, aluminium and steel.

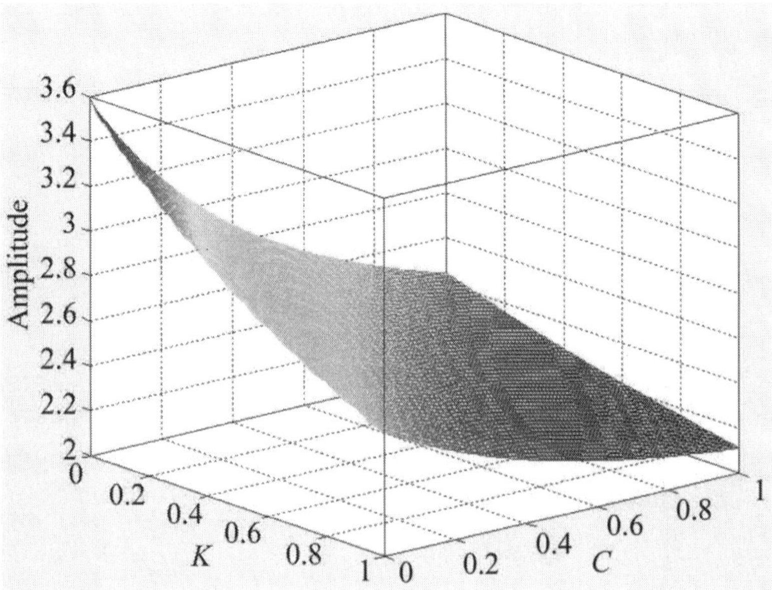

Figure 7. Effects of the viscoelastic coupling on the amplitude response of the system.

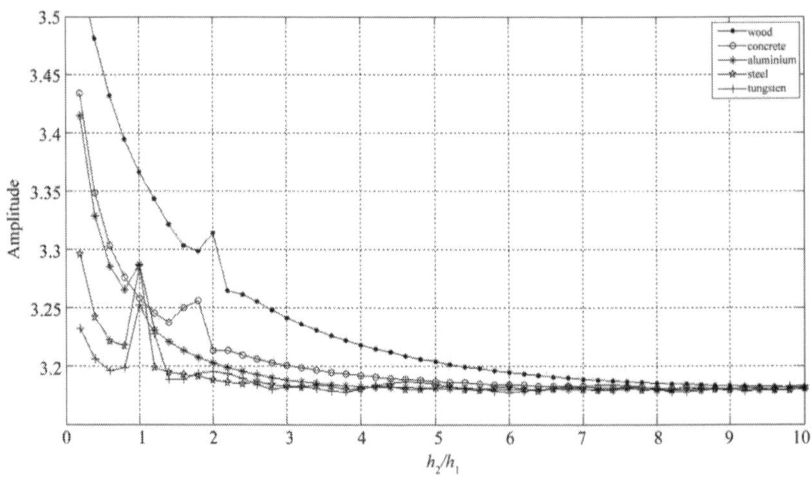

Figure 8. Maximal amplitudes of vibrations versus to the thickness of the dynamic absorbing plates with various characteristics.

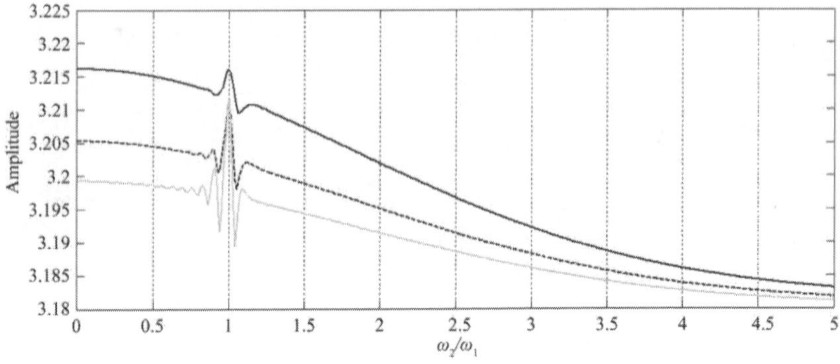

Figure 9. Maximal amplitudes of vibrations versus the frequency of the absorbing plate for $m_2 < m_1$ (continuous black line), $m_2 = m_1$(discontinuous black line), $m_2 > m_1$ (gray line).

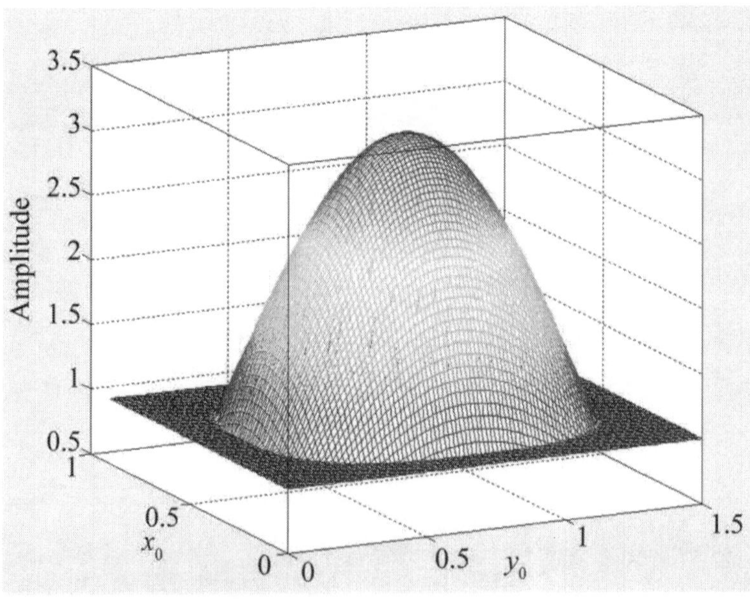

Figure 10. Amplitude of vibrations versus the position of the external forces.

5. CONCLUSION

The viscoelastic control has been proved to reduce or suppress amplitude of vibration of a rectangular plate under a localized periodic impulsive excitation. We started by modelling the impact force as a rectangular pulse which consider in this form enabled to subsequently model the dynamics of the plate under control for uniform distributions of the viscoelastic coupling. The modal equations derived from this modelling have been solved analytically and verified numerically. Considering the condition for the efficiency of control technics, we found that the control is effective whatever the values of the control parameters chosen. Moreover, the effectiveness of the control increases as the values of the control parameters becomes higher. We have also shown that the frequency of vibration of the dynamic absorbing plate has a positive effect on the control when it is greater than the one of the main plate. Regarding the effect of external excitation location on the plate, we have found that this position has particular effects on the intensities of amplitude of vibration and could influence considerably the dynamics response of the system.

ACKNOWLEDGEMENTS

Part of this work was completed during a research visit of Dr Nana Nbendjo at the Steel and Composite Structures Team of the University of Kassel in Germany. He is grateful to TWAS (The Academy of Science for the Developing World) and DFG (German Research Foundation) for financial support.

REFERENCES

1. T. T. Song and W. F. Chen, "Active Structural Control: Theory and Practice," Wiley, New York, 1990.

2. C. R. Fuller, S. J. Eliot and P. A. Nelson, "Active Control of Vibration," Academic Press, London, 1997.

3. S. H. Kim, S. B. Choi, S. R. Hang and M. S. Han, "Vibration Control of Fexible Structure Using a Hybrid Mounts," International Journal of Mechanical Sciences, Vol. 46, No. 1, 2004, pp. 143-157.

4. L. Jezequel, "Active Control in Mechanical Engineering," Hermes, New Castle, 1995.

5. B. R. Nana Nbendjo and P. Woafo, "Modelling, Control by Sandwich of the Dynamics and Optimization in a Strongly Nonlinear Beam," Far East Journal of Dynamical Systems, Vol. 8, 2006, pp. 267-283.

6. T. Aida, K. Kawazoe and S. Toda, "Vibration Control of Plates by Plate-Type Dynamics Vibration Absorber," Journal of Vibration and Acoustics, Vol. 117, No. 3, 1995, pp. 332-338.

7. T. Aida, S. Toda, S. N. Ogowa and Y. Simada, "Vibration Control of Beams by Beam Type Dynamics Vibrations Absorber," ASCE Journal of Engineering Mechanics, Vol. 118, No. 2, 1992, pp. 163-175.

8. B. R. Nana Nbendjo and P. Woafo, "Modeling and Optimal Active Control with Delay of the Dynamics of a Strongly Nonlinear Beam," Journal of Advanced Research in Dynamical and Control Systems, Vol. 1, 2009, pp. 57-74.

9. C. A. Kitio Kwuimy, B. R. Nana Nbendjo and P. Woafo, "Optimization of Electromechanical Control of Beam Dynamics:

Analytical Method and Finite Differences Simulation," Journal of Sound and Vibration, Vol. 298, No. 1-2, 2006, pp. 180-193.

10. A. A. Nanha Djanan, B. R. Nana Nbendjo and P. Woafo, "Control of Vibration on a Hinged-Hinged Beam under a Non-Ideal Excitation Using RLC Circuit with Variable Capacitance," Nonlinear Dynamics, Vol. 63, No. 3, 2011, pp. 477-489.

11. J. L. P. Felix and J. M. Balthazar, "Comments on a Nonlinear and Non-Ideal Electromechanical Damping Vibration Absorber, Sommerfeld Effect and Energy Transfer," Nonlinear Dynamics, Vol. 55, No. 1-2, 2009, pp. 1-11.

12. R. A. Morgan and R. W. Wang, "An Active Passive Piezoelectric Absorber for Structural Vibration Control under Harmonic Excitations with Time-Varying Frequency, part1: Algorithm Development and Analysis," Journal of Vibration and Acoustics, Vol. 124, No. 1, 2002, pp. 77-83.

13. M. S. Tsai and K. W. Wang, "On the Structural Damping Characteristics of Active Piezoelectric Actuator with Passive Shunt," Journal of Sound and Vibration, Vol. 221, No. 1, 1999, pp. 1-22.

14. I. Bica, "Damper with Magnetorheological Suspension," Journal of Magnetism and Magnetic Materials, Vol. 241, No. 2-3, 2002, pp. 196-200.

15. B. Liu and H. S. Tzou, "Distributed Photostrictive Actuator and Optopiezothermoelasticity Applied to Vibration Control of Plates," Journal of Vibration and Acoustics, Vol. 120, No. 4, 1998, pp. 936-943.

16. H.-R. Shih, H.-S. Tzou and M. Saypuri, "Structural Vibration Control Using Spacially, Configured Opto-Electromechanical Actuators," Journal of Sound and Vibration, Vol. 284, No. 1-2, 2005, pp. 361-378.

17. Q. Zhang and B. Z. Guo, "Stabilization of an Elastic Plate with Viscoelastic Boundary Conditions," Journal of Optimization Theory and Application, Vol. 122, No. 3, 2004, pp. 669-690.

18. H. P. Niu, Y. H. Zhang, X. H. Zhang and S. L. Xie, "Active Vibration Control of Plates Using Electro-Magnetic Constrained Layer Damping," International Journal of Applied Electromagnetics and Mechanics, Vol. 33, No. 1, 2010, pp. 831-837.

19. S. Lenci and G. Rega, "Periodic Solutions and Bifurcations in an Impact Inverted Pendulum under Impulsive Excitation," Chaos, Solitons and Fractals, Vol. 11, No. 15, 2000, pp. 2453-2472.

20. D. Zwillinger, "Handbook of Differential Equations," 3rd Edition, Academic Press, Cambridge, 1997.

CHAPTER 8

Shape Control and Vibration Analysis of Piezolaminated Plates Subjected to Electro-Mechanical Loading

Rajan L. Wankhade[1], Kamal M. Bajoria[2]

[1]Applied Mechanics Department, Government College of Engineering, Karad, India

[2] Department of Civil Engineering, Indian Institute of Technology Bombay, Mumbai, India

ABSTRACT

Shape control and free vibration analysis of piezolaminated plates subjected to electro mechanical loading are evaluated using finite element method. First order shear deformation theory is employed in the analysis. Both extensions as well as shear actuators are considered for piezolaminated plates. Rectangular four node isoparametric element is used in the finite element formulation. Variation of temperature is neglected for the orthotropic layers of the laminate and for piezolayer. Annular circular plates and rectangular plates with piezoelectric layers mounted and/or integrated are analysed for various parameters. Numerical results are presented for varying the actuator voltage for annular plates with different thicknesses of piezo patches. In case of rectangular plate shear actuator is considered for vibration analysis.

KEYWORDS

Piezoelectric Material, Finite Element Analysis, Shape control, Vibrational Analysis

1. INTRODUCTION

In recent years the use of smart piezolaminated composite structures has been increased. Due to their significant stiffness and low weight many isotropic materials and structures are now replaced by piezolaminated smart composites. The piezoelectric ceramics are available in the form of thin sheets which can be surface mounted and/or integrated with the laminates to control the structural response in shape, vibration and buckling. Suffieforts are directed towards the analysis of such piezoelectric laminated structures. It has gained an attraction toward the shape and vibration control of piezoelectric laminates which have the capability of active and passive control. Ha et al. (1992) performed finite element analysis of composite structures containing distributed piezoceramic sensors and actuators [1] . Hwang and Park (1993) studied finite element modeling of piezoelectric sensors and actuators and provided results for piezoelectric laminates [2] . Li and Bainum carried out vibration control analysis of flexible spacecraft integrating a momentum exchange controller and a distributed piezoelectric actuator [3]. Tzou and Zhou (1995) performed dynamics analysis and control of non-linear circular plates with piezoelectric actuators [4]. Thomson and Loughlan (1995) examined the active buckling control of some composite column using piezoceramic actuators [5]. Kim et al. (1996) developed finite element model for a smart cantilever plate and comparison also performed experimental analysis [6] . Batra and Liang (1997) investigated the vibrations of a rectangular laminated elastic plate with embedded piezoelectric sensors and actuators [7] . Agrawal et al. (1997) performed analysis of adaptive antenna for shape control using piezoelectric actuators [8] . Soares et al. (1999) carried out optimal design of piezolaminated structures [9] . Faria and Almeida (1999) studied enhancement of pre-buckling behavior of composite beams with geometric imperfections using piezoelectric actuators [10] . Sze and Yao (2000) used a hybrid stress ANS solid-shell element and its generalization for smart structure modeling [11] . Wang et al. (2001) performed vibration control of smart piezoelectric composite plates [12] . Cen et al. (2002) developed a new 4-node quadrilateral FE model with variable electrical degrees of freedom for the analysis of piezoelectric laminated composite plates [13] .

Moita et al. (2004) studied active control of adaptive laminated structures with bonded piezoelectric sensors and actuators. [14] . Artel and Becker (2005) investigated coupled and uncoupled analyses of piezoelectric free-edge effect in laminated plates [15] . Oh and Lee (2006) carried out analysis for supersonic flutter suppression of piezolaminated cylindrical panels based on multifield layerwise theory [16] . Ray and Balaji (2007) performed active structural acoustic control of laminated cylindrical panels using smart damping treatment [17] . Periasamy (2008) performed shape control of composite structures with optimally placed piezoelectic patches [18] . Ly et al. (2011) developed modeling and studied characterization of piezoelectric cantilever bending sensor for energy harvesting [19] . Kamal M. Bajoria and Rajan L. Wankhade (2012) analyzed rectangular piezolaminated plates for free vibration using Finite Element Method [20] . Wankhade and Bajoria (2013) developed finite element modeling for free vibration and buckling analysis of piezolaminated plates using higher order shear deformation theory [21] [22] . Brighenti (2014) studied smart behaviour of layered plates through the use of auxetic materials [23] . Bajoria and Wankhade (2015) performed vibration control of cantilever piezolaminated beam with extension and shear mode piezo actuators [24] . Kerboua et al.

(2015) further carried out vibration control of beam using piezoelectric-based smart material [25] . Zhang et al. (2015) studied active vibration control of piezoelectric bonded smart structures using PID algorithm [26] . Bendine and Wankhade studied vibration control of FGM piezoelectric plate based on LQR genetic search where analysis was performed for different sets of power law exponent [27] .

Here shape control and vibration analysis of piezolaminated plates subjected to combined action of electrical and mechanical loading is carried out considering first order shear deformation theory. Finite element method as employed for the analysis purpose. An isoperimetric four noded rectangular element is used in the finite element formulation. Shape control of annular plate with piezo patched at optimal locations is performed for varying the actuator voltage. Again the thickness of piezolayer is varied and response of structure is studied. Further in case rectangular plate shear piezoelectric actuator is considered for the vibration analysis. Results are presented considering shape control and vibration piezolaminated plate for different parameters viz. actuator voltage, aspect ratio of plate and varying thickness of piezolayer.

2. MATHEMATICAL FORMULATION

2.1. Equilibrium and Incremental Equations

Virtual displacement principle is considered in the mathematical formulation to obtain equation s of equilibrium. Equilibrium between internal and external forces has to be satisfied. If $\{\Psi\}$ represents the vector of the sum of the internal and external forces. Then equilibrium can be written as follow:

$$\{\Psi\} = \{R\} - \{P\} \tag{1}$$

In above equation, {R} is the external forces due to imposed load and {P} represents the vector of internal resisting forces. For piezolaminated plates the equilibrium equation can be written as:

$$\{\Psi\} = \{R\} + \frac{1}{2}\int_V \{\varepsilon\}^{\mathrm{T}} \{\sigma\} \mathrm{d}V - \frac{1}{2}\int_{V_a} \left[E_a^p\right]^{\mathrm{T}} \{D_a\} \mathrm{d}V - \frac{1}{2}\int_{V_s} \left[E_s^p\right]^{T} \{D_s\} \mathrm{d}V + \int_V \{\varepsilon^N\}^{\mathrm{T}} \{\sigma_0\} \mathrm{d}V \tag{2}$$

where V, V_s and V_a represents the area of the entire structure, sensor layer and actuator layer respectively. For writing the work done by external forces due to the applied surface traction and applied electric charge on actuator, the equation for external work done {R} can be summarized as:

$$\{R\} = \int_A \{\underline{u}\}^{\mathrm{T}} \{\bar{\sigma}(x,y)\} \mathrm{d}A + \int_A \phi_a' \bar{q}_a^e (x,y) \mathrm{d}A \tag{3}$$

2.2. Finite Element Modeling

Finite element modeling is considered for the analysis purpose. Here displacement field considered in finite element formulation is based on first order shear deformation theory. Hence considering the effect of shear deformation, displacement field of first order shear deformation theory is written as follow:

$$u = u_0 + z\theta_x$$

$$v = v_0 + z\theta_y$$

$$w = w_0 \tag{4}$$

where, u, v and w are the displacement of any point in the plate domain in x, y and z direction respectively. u_0, v_0 and w_0 are the displacement of midpoint of normal. θ_x, θ_y are the rotations of normal at the middle plane in x and y direction about y and x axis respectively.

The strains associated with the displacement model for linear bending are given by

$$\varepsilon_x = \frac{\partial u}{\partial x};\ \varepsilon_y = \frac{\partial v}{\partial y};\ \varepsilon_z = \frac{\partial w}{\partial z};\ \gamma_{xy} = \frac{\partial u}{\partial y} + \frac{\partial v}{\partial x};\ \gamma_{xz} = \frac{\partial u}{\partial z} + \frac{\partial w}{\partial x};\ \gamma_{yz} = \frac{\partial w}{\partial y} + \frac{\partial v}{\partial z} \tag{5}$$

Figure 1 shows general piezolaminated plate with actuator and sensor patches at the top and bottom surface of the plate. Fiber can be oriented with reference to the horizontal axes and is modelled in finite element formulation.

2.2.1. Shell Element and Shape Functions

Four nodded isoparametric rectangular element is adopted as shown in Figure 2.

Shape functions are given as:

$$N_i = \frac{1}{4}(1+\xi\xi_i)(1+\eta\eta_i), \quad i=1,2,3,4$$

2.2.2. Displacement Field

The displacement field associated with the eight nodded rectangular element is given as:

$$u = \sum_{i=1}^{8} N_i u_i$$

$$v = \sum_{i=1}^{8} N_i v_i$$

$$w = \sum_{i=1}^{8} N_i w_i \tag{6}$$

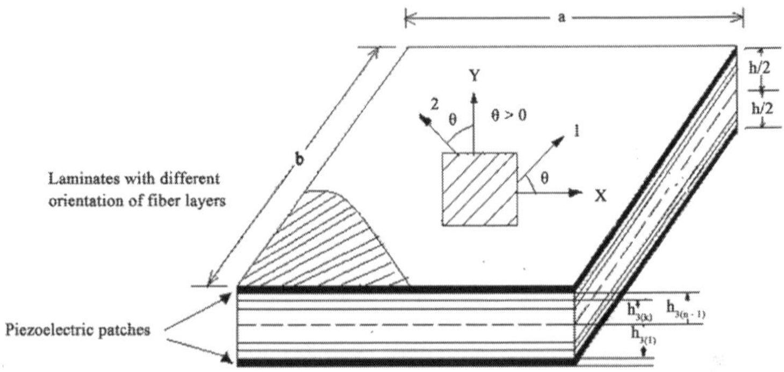

Figure 1. Piezolaminated plate with actuator and sensor layers at the top and bottom surface of the plate.

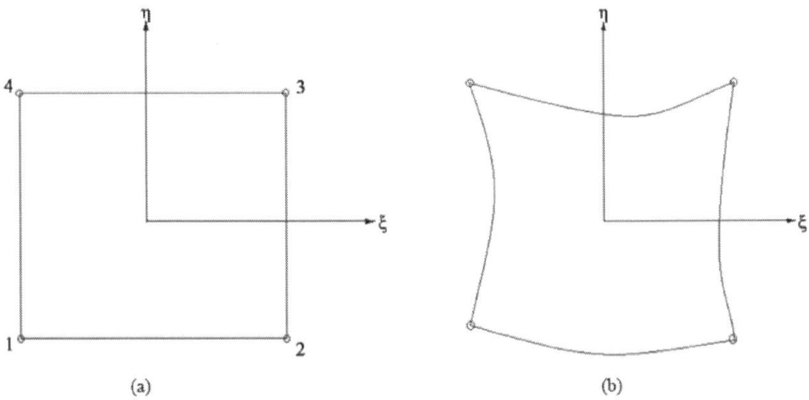

Figure 2. Four nodded isoparametric element. (a) Rectangular element; (b) Mapping of element.

2.2.3. Strain within the Element

Strains associated with the displacement field can be written as follow:

a. Middle plane membrane strains

$$\underbrace{\varepsilon_p}_{3\times 1} = \varepsilon_p^L + \varepsilon_p^N \tag{7}$$

in which ε_p^L and ε_p^N are linear and non-linear components of middle plane membrane strains and combining linear and non-linear terms membrane strains are given as:

$$\underbrace{\varepsilon_p}_{3\times 1} = \begin{Bmatrix} \dfrac{\partial u}{\partial x} \\ \dfrac{\partial v}{\partial y} \\ \dfrac{\partial u}{\partial x} + \dfrac{\partial v}{\partial y} \end{Bmatrix} + \begin{Bmatrix} \dfrac{1}{2}\left(\dfrac{\partial w}{\partial x}\right)^2 \\ \dfrac{1}{2}\left(\dfrac{\partial w}{\partial y}\right)^2 \\ \dfrac{\partial w}{\partial x} \cdot \dfrac{\partial w}{\partial y} \end{Bmatrix} \tag{8}$$

Curvature strains/Bending strains.

Curvature strains are linearly related to bending displacement as:

$$\varepsilon_b^L = \begin{Bmatrix} \varepsilon_{xxb} \\ \varepsilon_{yyb} \\ \varepsilon_{xyb} \end{Bmatrix} = z \begin{Bmatrix} \dfrac{\partial \theta_x}{\partial x} \\ \dfrac{\partial \theta_y}{\partial y} \\ \dfrac{\partial \theta_x}{\partial x} + \dfrac{\partial \theta_y}{\partial y} \end{Bmatrix} = z \cdot K \tag{9}$$

Shear strains,

$$\underbrace{\gamma}_{2\times1} = \begin{Bmatrix} \gamma_{xz} \\ \gamma_{yz} \end{Bmatrix} = \begin{Bmatrix} \theta_x + \dfrac{\partial w}{\partial x} \\ \theta_y + \dfrac{\partial w}{\partial y} \end{Bmatrix} \tag{10}$$

Thus

$$\left\{ \underbrace{\varepsilon}_{6\times1} \right\} = \begin{Bmatrix} \underbrace{\varepsilon_p}_{3\times1} \\ \underbrace{\varepsilon_b}_{3\times1} \end{Bmatrix} \tag{11}$$

and shear strains can separately be written as,

$$\underbrace{\gamma}_{2\times1} = \begin{Bmatrix} \gamma_{xz} \\ \gamma_{yz} \end{Bmatrix} \tag{12}$$

In which $\{\varepsilon_p\}, \{\varepsilon_b\}$ and $\{\gamma\}$ are membrane, bending and shear components of strains respectively. $\underbrace{\varepsilon}_{6\times1}$ is combined strain vector of membrane and bending strains. $\underbrace{\gamma}_{2\times1}$ is a vector containing shear strains. Subscript "p" stands for in-plane, "b" for bending, "L" for linear and subscript "N" stands for non-linear.

Hence strain displacement relationship can be obtained as:

$$\underbrace{\varepsilon} = B \underbrace{\delta_e} \tag{13}$$

2.3. Electro-Mechanical Coupling

For piezolaminated plates two constitutive relationships exist which includes the effect of mechanical and electrical loading as given by Eq. 17. Variation of temperature effect is neglected in formulation.

$$\{D\} = [e]\{\varepsilon\} + [g]\{E^p\}$$
$$\{\sigma\} = [C]\{\varepsilon\} - [e]^t\{E^p\} \tag{14}$$

In which,

$$\begin{Bmatrix} \sigma_{x'} \\ \sigma_{y'} \\ \sigma_{z'} \\ \sigma_{x'y'} \\ \sigma_{x'z'} \\ \sigma_{y'z'} \end{Bmatrix} = \begin{bmatrix} Q_{1'1'} & Q_{1'2'} & Q_{1'3'} & Q_{1'4'} & 0 & 0 \\ Q_{2'1'} & Q_{2'2'} & Q_{2'3'} & Q_{2'4'} & 0 & 0 \\ Q_{3'1'} & Q_{3'2'} & Q_{3'3'} & Q_{3'4'} & 0 & 0 \\ Q_{1'4'} & Q_{2'4'} & Q_{3'4'} & Q_{4'4'} & 0 & 0 \\ 0 & 0 & 0 & 0 & Q_{5'5'} & Q_{5'6'} \\ 0 & 0 & 0 & 0 & Q_{6'5'} & Q_{6'6'} \end{bmatrix} \begin{Bmatrix} \varepsilon_{x'} \\ \varepsilon_{y'} \\ \varepsilon_{z'} \\ \varepsilon_{x'y'} \\ \varepsilon_{x'z'} \\ \varepsilon_{y'z'} \end{Bmatrix} - \begin{bmatrix} 0 & 0 & e_{z1} \\ 0 & 0 & e_{z2} \\ 0 & 0 & e_{z3} \\ 0 & 0 & 0 \\ 0 & e_{y5} & 0 \\ e_{x6} & 0 & 0 \end{bmatrix} \begin{Bmatrix} E^p_{x'} \\ E^p_{y'} \\ E^p_{z'} \end{Bmatrix} \tag{15}$$

where, {D} is electric displacement vector, [e] is dielectric permittivity matrix, ε is the strain vector, $\{g\}$ is the dielectric matrix. $\{E\}$ is the electric field vector, $[\sigma]$ is the stress vector and [C] is the elastic matrix for constant electric field.

Electrical Potential Function

ϕ'_a and ϕ'_s are the electric displacement at any point in the actuator and the sensor layers, respectively, the electrical potential functions in terms of the nodal potential vector are written as:

$$\phi'_a = [N_{pa}]\{\phi^e_a\}$$

$$\phi'_s = [N_{ps}]\{\phi^e_s\} \tag{16}$$

where, $[N_{pa}]$ and $[N_{ps}]$ are the shape function matrices for the actuator and sensor layers, respectively. $\{\phi_a^e\}$ and $\{\phi_s^e\}$ are the nodal electric potential vector for the actuator and sensor layers, respectively.

2.4. Stiffness Matrix Equations

Element stiffness matrix is written as,

$$[K^e]\{\delta^e\}+[K_\sigma^e]\{\delta^e\}=[F_1^e]+[F_{ac}^e] \tag{17}$$

where,

$$[K^e]=[K_d^e]+[K_{aa}^e]^{-1}[K_{aa}^e][K_{ad}^e]+[K_{ds}^e][K_{ss}^e]^{-1}[K_{sd}^e] \tag{18}$$

$$\{F_{ac}^e\}=[K_{da}^e][K_{aa}^e]^{-1}\{Q_a^e\} \tag{19}$$

In which,

$$[K_d^e]=\int_V [B]^{\mathrm{T}}[C][B]\mathrm{d}V$$
,

$$[K_{da}^e]=[K_{ad}^e]^{\mathrm{T}}=\int_{Va}[B]^{\mathrm{T}}[e][B_a]\mathrm{d}V$$
,

$$[K_{aa}^e]=\int_{Va}[B_a]^{\mathrm{T}}[g][B_a]\mathrm{d}V$$
,

$$[K_{ds}^e]=[K_{sd}^e]^{\mathrm{T}}=\int_{Vs}[B]^{\mathrm{T}}[e][B_s]\mathrm{d}V$$

and

$$\left[K_{ss}^{e}\right]=\int_{Vs}[B_s]^{\mathrm{T}}[g][B_s]\mathrm{d}V \tag{20}$$

And hence element equation in the form of global stiffness matrix can be written as:

$$[K]\{\delta\}+[K_\sigma]\{\delta\}=[F_1]+[F_{ac}] \tag{21}$$

2.5. Free Vibration Equation

For free vibration problems, the equations of motion can be expressed as the following eigenvalue problem

$$\left([K]-\omega^2[M]\right)\{\delta_e\}=0 \tag{22}$$

where, the lowest magnitude of eigenvalue gives critical buckling load and the vector $\{d\delta\}$ represents the buckled mode shape. And hence the eigenvalue problem is solved so as to get the buckling loads.

3. RESULTS AND DISCUSSIONS

3.1. Shape Control of Annular Plate with Piezoelectric Patches

A shape control of clamped annular plate having diameter 1 m is analyzed subjected to electrical load. Here no mechanical load is provided to the structure. Figure 3 shows geometry of the clamped annular plate with piezoelectric patches provided on the surface of the plate. These piezoelectric patched are attached at specified location of plate Periasamy (2008) has analyzed the same annular plate with actuator voltage 45 only having constant tplate/tpiezo ratio [18] . Here the analysis of annular plate is carried out considering different actuator voltage of piezolaminated actuator. Plate is modelled with 4-noded rectangular element with different elements and results are shown inTable 1 and Table 2. It is found that 180 elements are sufficient to satisfy convergence criteria.

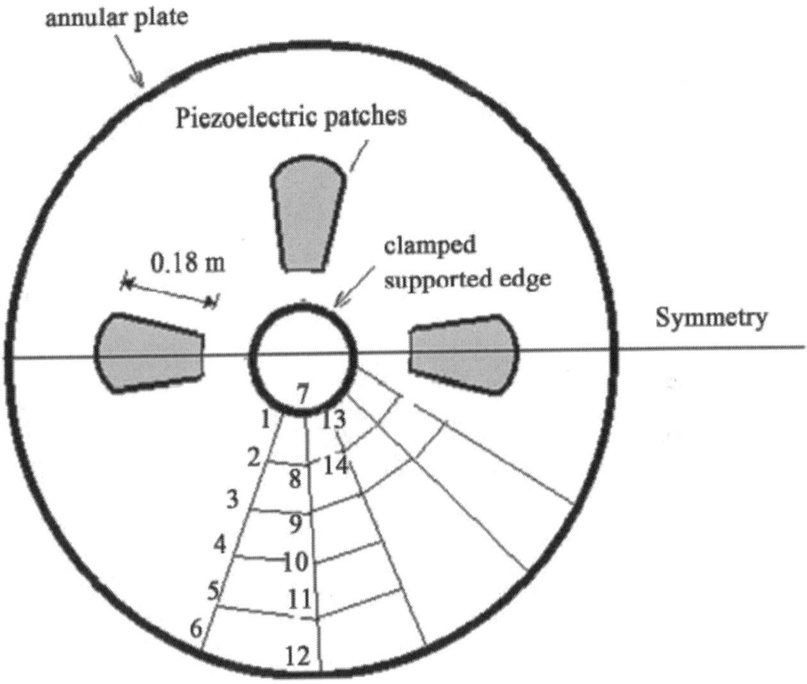

Figure 3. Geometry and modeling of clamped annular plate with piezoelectric patches.

Hence a convergence study is included to show that these numbers of elements are sufficient to get the realistic results and no need to go for higher number of elements. Figures 4-8 show deflection of the plate along its radius with varying actuator voltage (AV 20, AV 40 and AV 60) for different T_{plate}/T_{piezo} ratio as 1.5, 2. Material properties are adopted as follows:

Table 1. Convergence results with different plate element for a plate ($T_{plate}/T_{piezo} = 1.5$).

Sr. No.	No. of Elements	No. of Nodes	Deflection at Free Edge		
			AV 20	AV 40	AV 60
1	180	216	3.09E−04	6.17E−04	9.26E−04
2	720	792	3.00E−04	5.88E−04	9.21E−04
3	2880	3024	2.98E−04	5.86E−04	9.20E−04

Table 2. Convergence results with different plate element for a plate ($T_{plate}/T_{piezo} = 2.0$).

Sr. No.	No. of Elements	No. of Nodes	Deflection at Free Edge ($T_{plate}/T_{piezo} = 2.0$)		
			AV 20	AV 40	AV 60
1	180	216	2.31E−04	4.63E−04	6.94E−04
2	720	792	2.24E−04	4.49E−04	6.73E−04
3	2880	3024	2.22E−04	4.46E−04	6.68E−04

For graphite epoxy composite

$$E_{11}=150\text{ Gpa},\ E_{22}=9\text{ Gpa},\ G_{12}=G_{13}=7.1\text{ Gpa},G_{23}=2.5\text{ Gpa},\ \mu=0.3,\ \rho=1600\text{ kg/m}^3$$

For PZT G-1195

$$E_{11}=E_{22}=63\text{ Gpa},\ G_{12}=G_{13}=G_{23}=24.8\text{ Gpa},\ \mu=0.29,\rho=7600\text{ kg/m}^3,d_{31}=d_{32}=-166\text{pm/V},$$

The maximum tip displacements observed at end are 3.09E−4 and 6.17E−4 and 9.26E−4 for AV 20, AV 40 and AV 60 respectively with T_{plate}/T_{piezo} = 1.5. It is observed that increase in actuator voltage leads to considerable increase in end deflection. About 50% increase in deflection at the tip is found for AV40 as compared to that of AV 20. Also if we consider the thickness of plate to thickness of piezolayer ratio, about 57.2% increase in deflection is observed at free end as tplate to tpiezolayer ratio increases from 1.5 to 2.0. Thus either increase voltage or increase thickness of piezolayer it will have influence on shape control of structure.

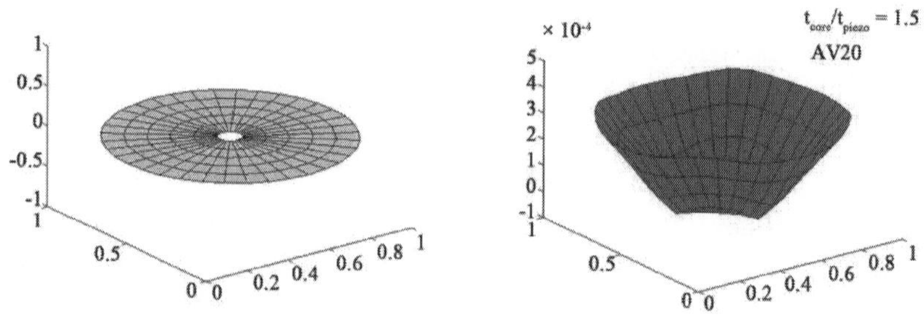

Figure 4. Undeformed and deformed shape of plate with actuator voltage 20 (T_{plate}/T_{piezo} = 1.5).

3.2. Free Vibration Analysis of Piezolamonated Plate with PZT Shear Actuator

A simply supported piezolaminated plate with the top and bottom layers made of 0^0 graphite-epoxy and the central layer made of axially poled PZT-5A thus having laminated scheme as [0^0 Gr-Epoxy/PZT-

5A/0^0 Gr-Epoxy] is examined for free vibration. Length of the plate is taken as 0.25 m.Table 3 shows elastic and piezoelectric properties of the laminate. The natural frequencies for different L/H ratio is presented in table 4. Total thickness of laminate is taken as "H", whereas the thickness of graphite-epoxy and that of PZT-5A are taken as 0.4H and 0.2H respectively.

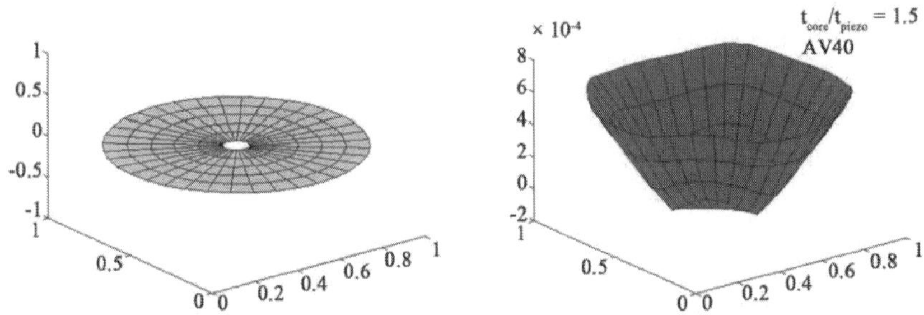

Figure 5. Undeformed and deformed shape of plate with actuator voltage 40 (T_{plate}/T_{piezo} = 1.5).

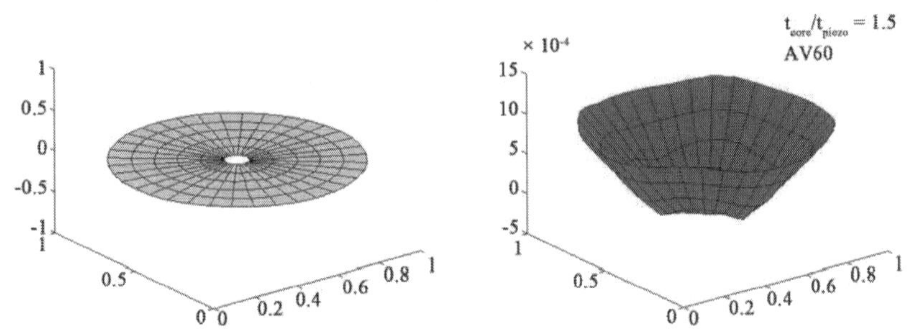

Figure 6. Undeformed and deformed shape of plate with actuator voltage 60 (T_{plate}/T_{piezo} = 1.5).

Table 4 shows natural frequencies in Hz for [Gr-epoxy/PZT-5A/ Gr-epoxy] simply supported shear actuated piezolaminated plate with closed loop and open loop electric condition. Maximum variation in frequency is found as 2.85% for closed loop as that of open loop.

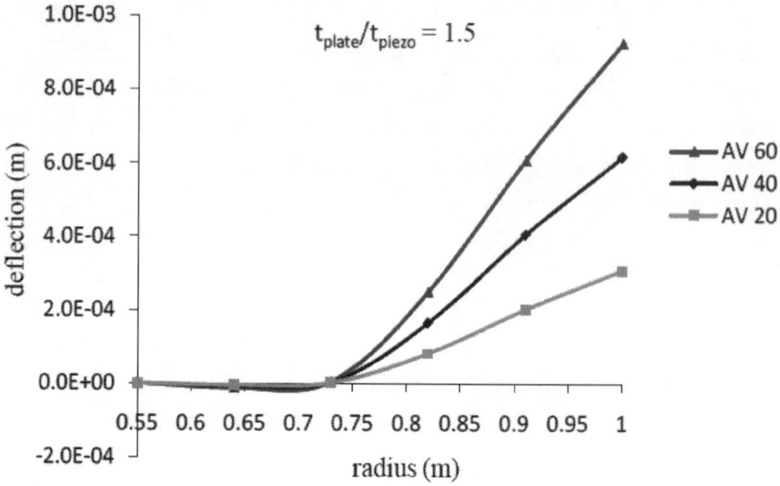

Figure 7. Deflection along the radius with varying actuator voltage (T_{plate}/T_{piezo} = 1.5).

Vibration analysis of piezoelectric laminated plate is carried out for different aspect ratio of plate. Piezolaminated plates are analysed for different electric condition of the piezolayer.

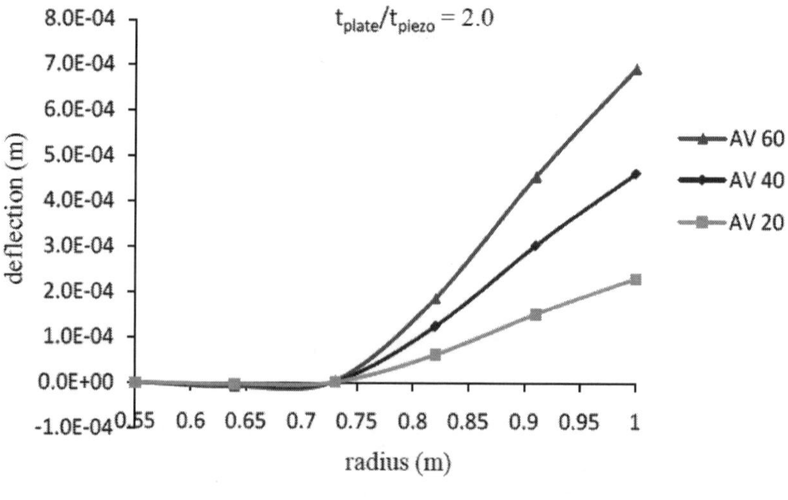

Figure 8. Deflection along the radius with varying actuator voltage (T_{plate}/T_{piezo} = 2.0).

Table 3. Elastic and piezoelectric properties for Graphite-Epoxy and PZT-5A.

Elastic Properties			Piezoelectric Properties		
Properties	Graphite-Epoxy	PZT-4	Properties	Graphite-Epoxy	PZT-5A
E_{11} (Gpa)	183.44	86.86	e_{11} (C/m^2)	0	−7.21
E_{22} (Gpa)	11.66	99.20	e_{12} (C/m^2)	0	−7.21
E_{33} (Gpa)	11.66	99.20	e_{13} (C/m^2)	0	12.32
G_{12} (Gpa)	4.363	50.78	e_{36} (C/m^2)	0	12.32
G_{23} (Gpa)	3.920	54.02	e_{35} (C/m^2)	0	12.32
G_{31} (Gpa)	4.363	50.78	g_{11} $(10^{-10}$ F/m)	153	150
υ_{12}	0.24	0.329	g_{22} $(10^{-10}$ F/m)	153	153
υ_{13}	0.24	0.432	g_{33} $(10^{-10}$ F/m)	153	153
υ_{23}	0.49	0.432			

Table 4. Natural frequencies (Hz) of [Gr-epoxy/PZT-5A/Gr-epoxy] simply supported shear actuated piezolaminated plate.

Mode	Aspect Ratio	Frequency	
		Closed Loop	Open Loop
1	4	2450.10	2482.38
	10	1327.86	1333.19
	100	144.014	144.71
2	4	6083.27	6217.93
	10	4364.49	4414.26
	100	1293.75	1295.16
3	4	9677.48	9930.18
	10	7931.00	8059.25
	100	3542.73	3547.78
4	4	13236.21	13624.61
	10	11578.50	11807.73
	100	5052.28	5060.17

4. CONCLUSION

In the present analysis, shape control and vibration of piezolaminated plates subjected to electro mechanical loading are studied based on finite element method. First order shear deformation theory is used for the same purpose. Both extensions as well as shear actuators are considered for two problems of piezolaminated plates. Annular circular plate and rectangular plates with piezoelectric layers mounted and/or integrated are analysed for various parameters. Numerical results are presented for varying the actuator voltage for annular plates with different thicknesses of piezo patches. In case of rectangular plate shear actuator is considered for vibration analysis. The maximum tip displacements observed at end are 3.09E−4 and 6.17E−4 and 9.26E−4 for AV 20, AV 40 and AV 60 respectively with $T_{plate}/T_{piezo} = 1.5$. Again the maximum tip displacements observed at end are 2.31E−4 and 4.63E−4 and 6.94E−4 for AV 20, AV 40 and AV 60 respectively with $T_{plate}/T_{piezo} = 2.0$. It is observed that increase in actuator voltage leads to considerable increase in end deflection. About 50% increase in deflection at the tip is found for AV40 as compared to that of AV 20 for same thickness of piezolayer. Also if we consider same voltage then deflection reduces for increasing thickness rations of plate to that of piezolayer. It is observed that with proper selection and placement of piezoelectric actuators as extension or shear, it is possible to generate enough forces on a structure in order to control its response in shape and vibration.

REFERENCES

1. Ha, S.K., Keilers, C. and Chang, F.K. (1992) Finite Element Analysis of Composite Structures Containing Distributed Piezoceramic Sensors and Actuators. AIAA Journal, 30, 772-780.

2. Hwang, W.S. and Park, H.C. (1993) Finite Element Modeling of Piezoelectric Sensors and Actuators. AIAA Journal, 31, 930-937.

3. Li, Z. and Bainum, P.M. (1994) Vibration Control of Flexible Spacecraft Integrating a Momentum Exchange Controller and a Distributed Piezoelectric Actuator. Journal of Sound and Vibration, 177, 539-553.

4. Tzou, H.S. and Zhou, Y.H. (1995) Dynamics and Control of Non-Linear Circular Plates with Piezoelectric Actuators. Journal of Sound and Vibration, 188, 189-207.

5. Thomson, S.P. and Loughlan, J. (1995) The Active Buckling Control of Some Composite Column Using Piezoceramic Actuators. Composite Structures, 32, 59-67.

6. Kim, J., Varadan, V.V., Varadan, V.K. and Bao, X.Q. (1996) Finite Element Modeling of a Smart Cantilever Plate and Comparison with Experiments. Smart Materials and Structures, 5, 165-170.

7. Batra, R.C. and Liang, X.Q. (1997) The Vibration of a Rectangular Laminated Elastic Plate with Embedded Piezoelectric Sensors and Actuators. Computers and Structures, 63, 203-216.

8. Agrawal, B.N., Elshafei, M.A. and Song, G. (1997) Adaptive Antenna Shape Control Using Piezoelectric Actuators. Acru Astronautica, 40, 821-826.

9. Soares, C.M.M. and Correia, V.M.F. (1999) Optimal Design of Piezolaminated Structures. Composite Structures, 47, 625-634.

10. Faria, A.R. and Almeida, S.F.M. (1999) Enhancement of Pre-Buckling Behavior of Composite Beams with Geometric Imperfections Using Piezoelectric Actuators. Composites Part B, 30, 43-50.

11. Sze, K.Y. and Yao, L.Q. (2000) A Hybrid Stress ANS Solid-Shell Element and Its Generalization for Smart Structure Modeling: Part I Solid Shell Element Formulation. International Journal for Numerical Methods in Engineering, 48, 545-564.

12. Wang, S.Y., Quek, S.T. and Ang, K.K. (2001) Vibration Control of Smart Piezoelectric Composite Plates. Smart Materials and Structures, 10, 637-644.

13. Cen, S., Soh, A.K., Long, Y.Q. and Yao, Z.H. (2002) A New 4-Node Quadrilateral FE Model with Variable Electrical Degrees of Freedom for the Analysis of Piezoelectric Laminated Composite Plates. Composite Structures, 58, 583-599.

14. Moita, J.M.S., Correia, I.F.P., Soares, C.M. and Soares, C.A.M. (2004) Active Control of Adaptive Laminated Structures with Bonded Piezoelectric Sensors and Actuators. Computers and Structures, 82, 1349-1358.

15. Artel, J. and Becker, W. (2005) Coupled and Uncoupled Analyses of Piezoelectric Free-Edge Effect in Laminated Plates. Composite Structures, 69, 329-335.

16. Oh, I.K. and Lee, I. (2006) Supersonic Flutter Suppression of Piezolaminated Cylindrical Panels Based on Multifield Layerwise Theory. Journal of Sound and Vibration, 291, 1186-1201.

17. Ray, M.C. and Balaji, R. (2007) Active Structural Acoustic Control of Laminated Cylindrical Panels Using Smart Damping Treatment. International Journal of Mechanical Sciences, 49, 1001-1017.

18. Periasamy, R. (2008) Shape Control of Composite Structures with Optimally Placed Piezoelectic Patches. Ms Thesis, University of Waterloo, Ontario.

19. Ly, R., Rguiti, M., D'Astorg, S., Hajjaji, A., Courtois, C. and Leriche, A. (2001) Modeling and Characterization of Piezoelectric Cantilever Bending Sensor for Energy Harvesting. Sensors and Actuators A: Physical, 168, 95-100.

20. Bajoria, K.M. and Wankhade, R.L. (2012) Free Vibration of Simply Supported Piezolaminated Composite Plates Using Finite Element Method. Advanced Materials Research, 587, 52-56.

21. Wankhade, R.L. and Bajoria, K.M. (2013) Buckling Analysis of Piezolaminated Plates Using Higher Order Shear Deformation Theory. International Journal of Composite Materials, 3, 92-99.

22. Wankhade, R.L. and Bajoria, K.M. (2013) Free Vibration and Stability Analysis of Piezolaminated Plates Using Finite Element Method. Smart Materials and Structures, 22, Article ID: 125040.

23. Brighenti, R. (2014) Smart Behaviour of Layered Plates through the Use of Auxetic Materials. Thin walled Structures, 84, 432-442.

24. Bajoria, K.M. and Wankhade, R.L. (2015) Vibration of Cantilever Piezolaminated Beam with Extension and Shear Mode Piezo Actuators. Proceedings of SPIE, Active and Passive Smart Structures and Integrated Systems, 9431, 943122.

25. Kerboua, M., Megnounif, A., Benguediab, M., Benrahou, K.H. and Kaoulala, F. (2015) Vibration Control Beam Using Piezoelectric-Based Smart Materials. Composite Structures, 123, 430-442.

26. Zhang, S., Schmidt, R. and Qin, X. (2015) Active Vibration Control of Piezoelectric Bonded Smart Structures Using PID Algorithm. Chinese Journal of Aeronautics, 28, 305-313.

27. Bendine, K. and Wankhade, R.L. (2016) Vibration Control of FGM Piezoelectric Plate Based on LQR Genetic Search. Open Journal of Civil Engineering, 6, 1-7.

CHAPTER 9

Adaptive Piezoelectric Absorber for Active Vibration Control

Sven Herold * and Dirk Mayer

Fraunhofer Institute for Structural Durability and System Reliability LBF, Bartningstr. 47, 64823 Darmstadt, Germany

ABSTRACT

Passive vibration control solutions are often limited to working reliably at one design point. Especially applied to lightweight structures, which tend to have unwanted vibration, active vibration control approaches can outperform passive solutions. To generate dynamic forces in a narrow frequency band, passive single-degree-of-freedom oscillators are frequently used as vibration absorbers and neutralizers. In order to respond to changes in system properties and/or the frequency of excitation forces, in this work, adaptive vibration compensation by a tunable piezoelectric vibration absorber is investigated. A special design containing piezoelectric stack actuators is used to cover a large tuning range for the natural frequency of the adaptive vibration absorber, while also the utilization as an active dynamic inertial mass actuator for active control concepts is possible, which can help to implement a broadband vibration control system. An analytical model is set up to derive general design rules for the system. An absorber prototype is set up and validated experimentally for both use cases of an adaptive vibration absorber and inertial mass actuator. Finally, the adaptive vibration control system is installed and tested with a basic truss structure in the laboratory, using both the possibility to adjust the properties of the absorber and active control.

KEYWORDS

inertial mass actuator; piezoelectric actuator; adaptive system; vibration control

1. INTRODUCTION

Tuned vibration absorbers (TVA) have been used since the beginning of the 20th century to reduce disturbing vibrations [1]. The original development aimed at damping of resonant vibrations with a damped absorber system. To cancel narrowband, forced vibrations, the TVA can be tuned to the excitation frequency, as well. In this case, the device is just lightly damped and often referred to as a *vibration neutralizer* in the literature [2]. To achieve sufficient performance over a wider bandwidth, the mass of the vibration neutralizer can be increased; alternatively, the vibration neutralizer can be designed to adapt its resonance frequency and to track the excitation frequency. Several basic concepts exist to adjust the resonance by actuation. Overviews can be found in [2,3,4]; also, some concepts will be introduced in the following section.

In previous studies, adaptive absorber concepts were evaluated regarding their performance, range of adaptation and reliability in fatigue tests [5]. It turned out that concepts that utilize piezoelectric actuation for the adaptation are more reliable compared to mechatronic solutions with servo motors due to less parts and no movable parts being used. Furthermore, the TVA can be dynamically excited above its resonance frequency and work as an inertial mass actuator [6], enhancing the bandwidth of the vibration reduction [7].

The work presented in this paper deals with investigations on a dynamic inertial mass actuator that has been developed previously [8] and which can also be used as an adaptive tuned vibration absorber. In the following section, the basic concepts for the adaptation of the resonance frequency are summarized and compared. The design of the adaptive absorber that will be investigated is introduced. Experiments on the performance of the different functions of the device are performed at a truss structure that was already used as an example for other laboratory experiments [8,9]. The structure can be considered a generic example of lightweight structures used in space applications or bridges. Similar experimental set ups have already been implemented and used for the evaluation of vibration control systems [10].

2. ADAPTIVE ABSORBERS AND NEUTRALIZERS

The general idea of a vibration neutralizer is to generate counteracting forces on a system that is excited by narrowband disturbing forces. A generic example for the system set up is shown in Figure 1.

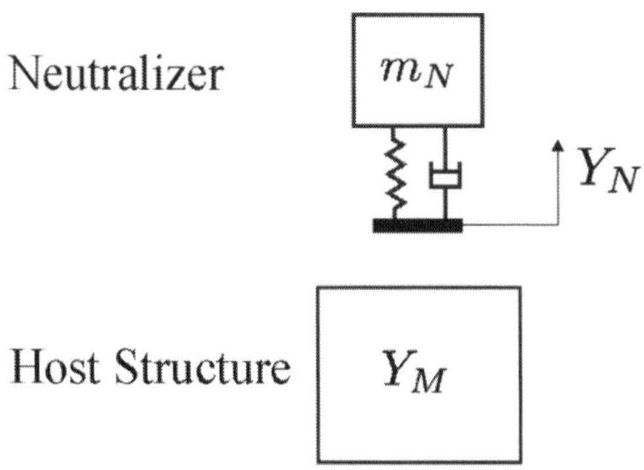

Figure 1. Generic example for a system with a vibration neutralizer.

The resulting frequency response function can be calculated using mechanical impedance formulations [9]. The admittance formulation for the neutralizer, which can be characterized by the mass *m*, resonance frequency ω0ω0 and damping coefficient θ, reads:

$$Y_N(s) = \frac{1}{m}\frac{s^2 + s2\omega_0\theta + \omega_0^2}{s^2 2\omega_0\theta + s\omega_0^2} \quad (1)$$

With input admittance of the host structure YMYM, the admittance of the coupled subsystems results:

$$Y(s) = \frac{1}{m}\frac{s^2 + s2\omega_0\theta + \omega_0^2}{s^2 + s2\omega_0\theta + \omega_0^2 + Y_M(s)m_N(s^2 2\omega_0\theta + s\omega_0^2)} \quad (2)$$

Obviously, the neutralizer causes two conjugate complex zeros in the resulting admittance, but also an additional pair of poles. A simple example with a mass of 1kg1kg representing the host structure and an excitation force of 0.5N0.5N illustrates the resulting admittance characteristics (Figure 2). For a neutralizer of 0.1kg0.1kg tuned to 50Hz50Hz, a significant drop in the vibration amplitude at the resonance is observed, but also a peak at about 53Hz. If the excitation frequency were to increase above 50Hz towards this frequency, the neutralizer would cause a deterioration of the vibrations. If a much larger neutralizer mass of 0.5kg is used, the bandwidth of vibration reduction is increased; however, in most practical problems, it is not possible to add another 50% to the host's mass.

This motivates the integration of active elements to adjust the resonance frequency of the neutralizer. The bandwidth of the vibration reduction can be increased while still maintaining realistic mass ratios. Some basic concepts can be distinguished, here illustrated with the example of a simple vibration neutralizer based on a bending beam spring (Table 1).

The first option is to induce a pre-stress into the beam in order to change its effective bending stiffness. This requires a static force that has to be steadily applied when the stiffness should be tuned. The most straightforward option is to adjust the geometry, e.g., the effective length of the beam, by moving the inertial mass. The advantage of this concept is that a variety of actuators is available for such a task. However, the movable parts in the system are prone to non-linearities and are sources of mechanical failures. The third way to be introduced here is the utilization of dynamic feedback. A dynamic feedback force (in the picture indicated by a piezo patch transducer on the bending beam) that is proportional to the acceleration of the inertial mass can adjust the effective mass of the absorber. Due to the dynamic feedback loop, this concept requires highly dynamic actuation systems. Furthermore, the feedback loop might have certain stability limits, which restrict the tuning range of the resonance frequency: in most elastic systems, more than a single vibration mode exists. For the acceleration feedback, the stability of the control loop is only guaranteed if the sensor and the actuator are*collocated*, which includes the minimum-phase behavior of the open loop [11]. However, for higher order modes of the bending beam, the sensor and actuator are not collocated any more, which limits the stability of the controlled system. Still, the main advantage of this concept is that the system can be used as an inertial mass actuator for active control of vibrations at the same time using a second control loop. The mechanical set up is identical to the adaptive absorber (Figure 3), but if the actuator is driven well above the resonance frequency of the system, the device can be considered a nearly ideal force generator. In this case, the generated force equals the actuator force (Figure 4), alleviating the task of control system design for the active control of vibrations.

With the transverse actuator force being FT,AFT,A, the block force of the inertial mass actuator reads:

$$\frac{F}{F_{T,A}} = \frac{s^2}{s^2 + 2\theta\omega_0 s + \omega_0^2} \tag{3}$$

Obviously, all concepts introduced possess specific benefits and drawbacks. In this work, an adaptive vibration absorber concept is investigated that can also be used as an inertial mass actuator. However, the resonance is adapted by using static pre-stress forces instead of dynamic actuation. This should enable a wide tuning range for the resonance frequency.

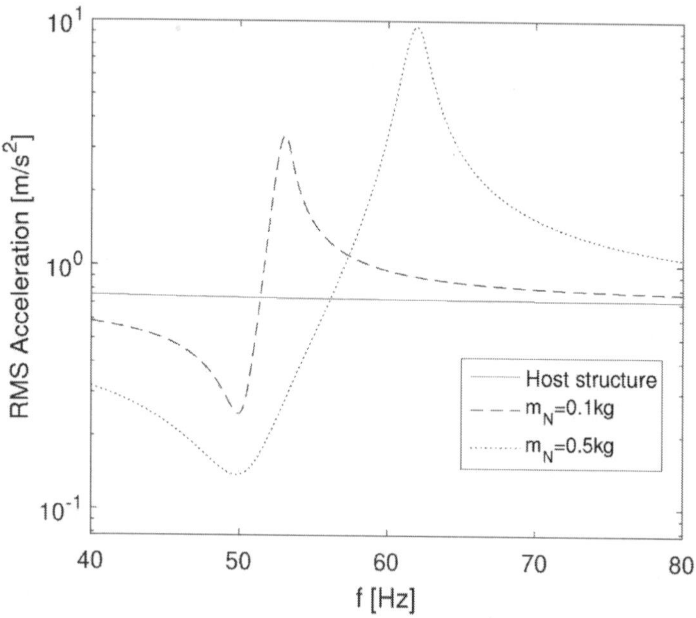

Figure 2. Vibration amplitude for a structure without and with neutralizers of different masses.

3. DESIGN OF THE ADAPTIVE PIEZOELECTRIC ABSORBER

The considerations are applied to design an adaptive vibration absorber with integrated piezoelectric transducers. A well-known design concept for vibration absorbers consists of a thin beam with a tip mass, which is used in this paper, as well. To activate the absorber, two piezoelectric stack transducers, working as actuators, are mounted in parallel to the beam (Figure 5).

Table 1. Different adaptation concepts for neutralizers [5].

Concept	Static Pre-Stress	Variable Geometry	Dynamic Forces
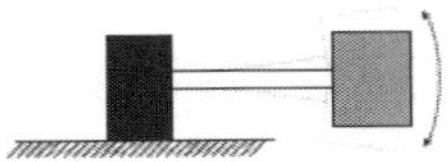	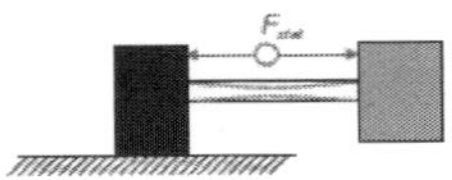	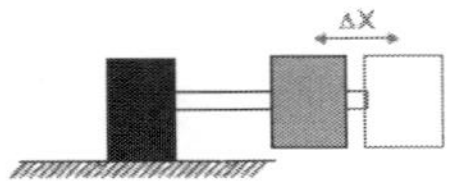	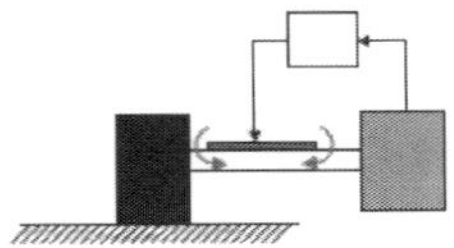
Actuator type	Static, high force	Static, high stroke	Dynamic force
Examples	Piezoelectric	Motor, shape memory alloys	Piezoelectric, electrodynamic
Energy Supply	Continuously	During adaptation	Continuously

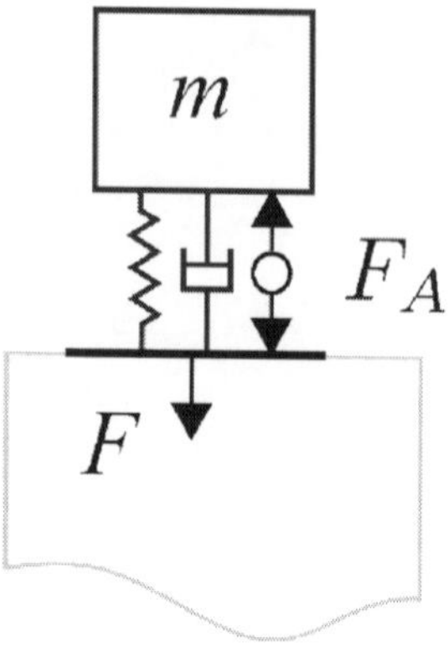

Figure 3. Principle of an inertial mass actuator.

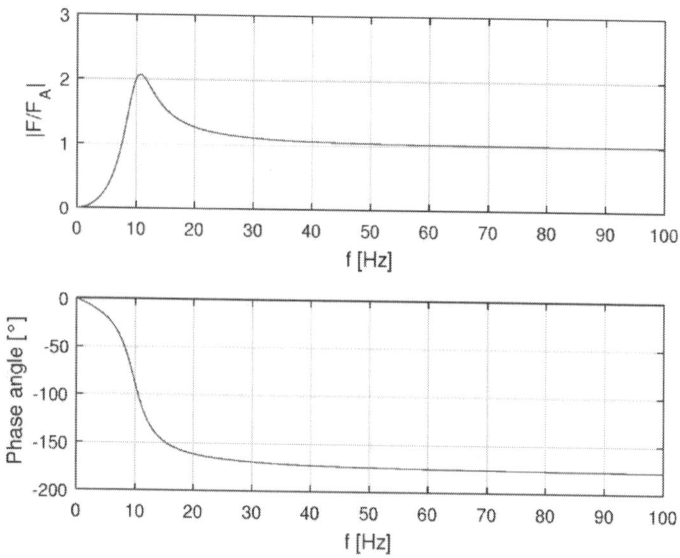

Figure 4. Frequency response function of an inertial mass actuator (m=1kg,f0=10Hz,θ=0.25m=1kg,f0=10Hz,θ=0.25).

The application of tensile normal force to the thin beam by the static elongation (in phase) of piezoelectric actuators enables the adaptation of the absorber's natural frequency. Alternatively, the actuators can be used in dynamic operation (out of phase) to generate transverse forces. Adaptive behavior in this operation mode is obtained introducing a virtual change of the mass or the stiffness of the absorber by acceleration, respectively displacement, feedback control.

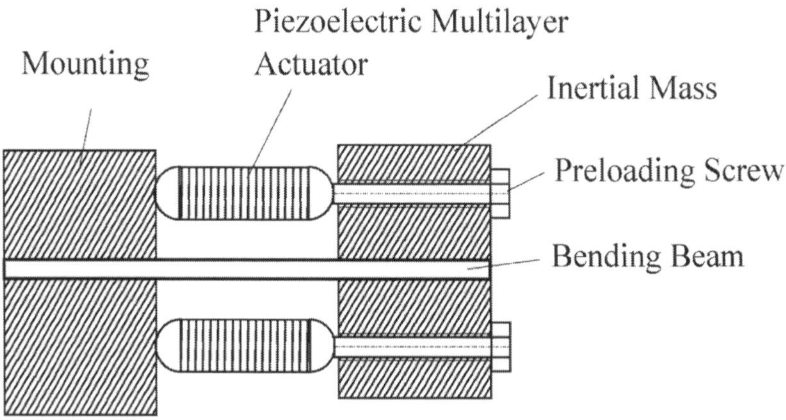

Figure 5. Basic design of the adaptive vibration absorber.

3.1. Basic Equations

The integration of the beam and the piezoelectric actuators is simplified in order to enable a modular, reconfigurable adaptive absorber system (Figure 5 and Figure 6). The two piezoelectric actuators are mounted at a certain distance to the neutral axis of the beam. The inertial mass is intended to serve also as a lever for the actuation.

Applying a static voltage to both actuators, a tensile (normal) force is induced at the beam, which is responsible for additional beam stiffness. In general, this effect can be described by the differential equation for the axially-loaded Euler-Bernoulli beam (4).

$$EIw'''' + F_N w'' = 0 \tag{4}$$

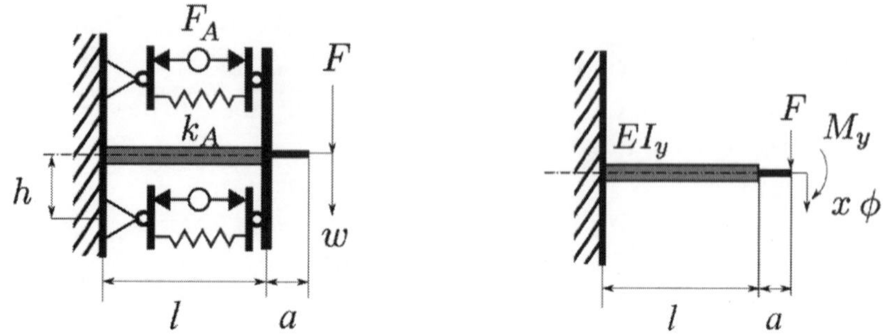

Figure 6. Mechanical sketch of the adaptive vibration absorber.

A solution of the aforementioned differential equation can be obtained by the trial function:

$$w = A_1 + A_2 l + A_3 \cos(\lambda l) + A_4 \sin(\lambda l) \quad \text{with} \quad \lambda = \sqrt{\frac{F_N}{EI}} \tag{5}$$

The coefficients can be found by solving a system of Equation (7) defined by the boundary conditions of the system:

$$\begin{array}{ll} w(x=0)=0 & EIw''(x=l)+2h(F_{N,B}-k_A h w'(x=l))-Fa=0 \\ w'(x=0)=0 & EIw'''(x=l)+F_N w'(x=l)+F=0 \end{array} \tag{6}$$

This leads to:

$$\mathbf{C}\cdot\mathbf{A}=\mathbf{F}$$

$$\mathbf{C}=\begin{bmatrix} 1 & 0 & 1 & 0 \\ 0 & 1 & 0 & \lambda \\ 0 & 2k_A h^2 & -\lambda^2 EI\cos(\lambda l)-2k_A h^2\lambda\sin(\lambda l) & -\lambda^2 EI\sin(\lambda l)+2k_A h^2\lambda\cos(\lambda l) \\ 0 & F_n & \lambda^3 EI\sin(\lambda l)-F_N\lambda\sin(\lambda l) & -\lambda^3 EI\cos(\lambda l)+F_N\lambda\cos(\lambda l) \end{bmatrix}$$
$$\mathbf{F}=\begin{bmatrix} 0 \\ 0 \\ 2hF_{N,B}+M+aF \\ -F \end{bmatrix} \tag{7}$$

In order to include the geometry of the mass, the distance from the end of the beam to the center of gravity of the mass is defined by *a* (refer to Figure 6). In Equation (8), the displacement w(x=l+a) is evaluated at this point.

$$w\left(x=l+a\right) \quad = \quad w\left(x=l\right)+a\,w'\left(x=l\right) \tag{8}$$

Transforming Equation (8) according to the external transverse load *F* and dividing the equation by this force, the effective transverse stiffness kt,effkt,eff at x=l+a of the system depending on the normal force in the beam is derived.

$$k_{t,eff}=\left(\frac{w\left(x=l+a\right)}{F}\right)^{-1} \tag{9}$$

For the investigation, generic piezoelectric actuators are assumed, which are described by their longitudinal stiffness kAkAand block force FN,B. Both actuators are coupled to the tip of the beam by a lever of length *h* with respect to the neutral axis of the beam. The beam itself is characterized by its length *l*, Young's modulus *E* and the moment of inertia IyIy. Since the actuators are coupled mechanically by spherical end caps, only normal forces are assumed to be transmitted at the mounting points of the actuators. External loads *F* and *M* are considered in order to couple the beam to other sub-systems, e.g., the tip mass.

By driving both actuators out of phase, a moment is generated at the tip of the beam, which causes a movement of the mass and, in turn, a transverse reaction force at the mounting. This mode of operation is preferably used for dynamic applications.

Tuning of the actuator by mechanical pre-loading is discussed in [8]. There numerical and experimental results show a good correlation. The static pre-loading by the piezoelectric actuators (electrical pre-loading) is more suitable. It allows for a simple adjusting of the TVA's resonance frequency during operation. A huge number of contributions discussing piezoelectric actuators have been published in the last decades ([12,13,14,15] and many more). A detailed simplification of the linearized piezoelectric equations for the one-dimensional case is derived, for example, in [16]. In this paper, the force generation of the one-dimensional actuator equation is of special interest:

$$F_A = \frac{E_A A_A}{l} u + \frac{d_{33} E_A A_A n_A}{l} U_{el} \tag{10}$$

The actuator force FAFA is defined by summarizing the mechanical and the electro-mechanical force. The parameters of the actuator are the Young's modulus EAEA, the cross-section AAAA, the length l, the electro-mechanical constant d33d33 and the number of layers nAnA. Mechanical displacement u and electrical voltage UelUel are responsible for generating the actuator force. If the displacement u is set to zero, the block force FN,B for the piezoelectric actuator can be derived (11).

$$F_{N,B} = \frac{d_{33} E_A A_A n_A}{l} U_{el} \tag{11}$$

However, if the actuators are elongating, driven by a voltage, the beam is pre-stressed. For this internal effect, the serial connection of both members defines the effective displacement of the actuator (12).

$$u_{eff} = \frac{F_{N,B}}{k_{N,eff}} \quad \text{with} \quad k_{N,eff} = \frac{2k_a k_{N,b}}{2k_a + k_{N,b}} \tag{12}$$

The effective stiffness kN,eff of the system is therefore defined by the serial arrangement of the actuator stiffness kAkA and the beam stiffness kN,bkN,b (refer to Equation 12). In this case, the normal force of the beam is identical to the force generated by both actuators. Using this relation, the bending stiffness kt,eff at x=l+a can be expressed with respect to the actuator voltage. In Section 4, the analytical and experimental results are compared, and a good match is observed.

3.2. Mechanical Design of the Adaptive Absorber

According to the model (Figure 5) a prototype was built that is shown in Figure 7.

Figure 7. Implemented prototype of the adaptive vibration absorber.

A thin beam is clamped between two steel blocks, which form the mounting of the inertial mass actuator. The width of the beam is large compared to its length to prevent torsion. The hole pattern of the mounting is suited for a connection to a shaker or a heavy breadboard. On the other side of the beam, the inertial mass is attached. It consists of an equal number of steel blocks on each side of the beam, which can be changed easily for experiments. The mass can be moved on the beam via slotted holes, which is necessary for the assembling of the piezoelectric actuators. These are monolithic multilayer actuators by CeramTec with a base area of 7.1mm7.1mm×7.6mm7.6mm and a length of 30mm30mm. On both sides of the piezoelectric actuators, spherical ceramic caps are glued to reduce damaging shear forces. To ensure the force transmission of the actuators and to avoid loosening and, therefore, the non-linear behavior, the actuators have to be pre-loaded. This is done by screws in the two masses next to the beam with a distance of 7.5mm7.5mm to the beam's surface. The tips of the pre-loading screws are concave as counterparts of the actuator's ceramic caps. There are corresponding holes at the mounting at the same distance from the beam. This prevents shifting of the piezoelectric actuators in operation.

4. EXPERIMENTAL VALIDATION

4.1. Analysis of the Adaptive Absorber Characteristics

To analyze the dynamic characteristics of the adaptive absorber, tests were performed. The parameters for the actuator system are depicted in Table 2. The whole actuator was mounted to a rigid, heavy base plate, which can be treated as an infinitely small mechanical admittance in the considered frequency range (Figure 8). The tip mass was instrumented with an accelerometer. In order to estimate the resonance frequencies and analyze the tuning capabilities of the adaptive absorber, frequency response functions (FRF) were measured using a dynamic signal analyzer (Ono Sokki CF5220). The excitation force was applied with an impulse hammer, and the acceleration of the mass was measured with the attached sensor. A DC voltage was applied to both piezoelectric actuators by two amplifiers SVR 150 (Piezomechanik) to generate the pre-stress force, which tunes the vibration absorber. FRFs for different DC voltages between 0V0V and 140V140V were measured.

The resulting FRFs show the tuning capabilities of the system (Figure 9). Furthermore, the analytical model introduced in the preceding section is validated. The predicted FRFs by the model match well with the experimental results. Some differences can be observed for very low DC voltages, where the absorber does not exhibit perfect characteristics of an single-degree-of-freedom oscillator. This could be caused by a poor mechanical coupling between the piezo stacks and the structure when no pre-load is applied. For very high pre-loads around 140V140V, the model predicts higher tuning effects than measured in the experiment (Figure 10). In this case, the analytical model might not perfectly predict the contact characteristics of the ball joints, which can be non-linear for high mechanical loads.

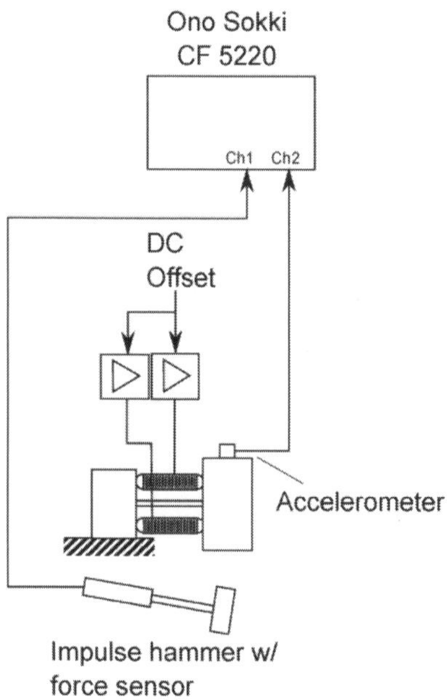

Figure 8. Test set up for the dynamic analysis of the adaptive vibration absorber.

Overall, tuning between 23Hz23Hz and 48Hz48Hz was possible by DC pre-loads. This is equivalent to an adaptive absorber with a center frequency of 35.5Hz35.5Hz and a tuning of ±35±35. This performance is comparable to adaptive absorbers, which use motors for the variation of the spring geometry investigated in preliminary work [5].

Table 2. Parameters of the adaptive vibration absorber.

Quantity	Symbol	Value	Unit
beam			
length	l	41.5	mm
width	b	60	mm
thickness	d	0.5	mm
Young's modulus	E	2.2×10^{11}	$\mathrm{N\,m^{-2}}$
mass			
mass	m	1.078	kg
actuator			
stiffness	k_A	42×10^{6}	$\mathrm{N\,m^{-1}}$
max. block force	$F_{B,max}$	2000	N

4.2. Inertial Mass Actuator Performance

For the evaluation of the generated active force of the device, the piezoelectric actuators were driven with a swept sine signal generated by the analyzer. A simple analogue circuit generates an out of phase driving signal for one of the actuators in order to excite the bending mode of the actuator system, and the acceleration of the tip mass was measured (Figure 11). Since the mounting plate can be assumed not to interact with the active system, the block force can be directly derived from the acceleration and the mass.

For this test, different pre-loads from 20 to 120VDC120VDC were applied to the actuators. The resulting FRFs between applied driving voltage and the generated force show the expected characteristics (Figure 12). In the frequency range above the first resonance and 250Hz250Hz, no further resonances are excited. However, it was observed that the FRFs are influenced by the pre-load. The block force above the resonance increases slightly with the pre-loading voltage. This might be caused by a non-linearity in the piezoelectric elements, *i.e.*, a dependency of the piezoelectric constant to the pre-load.

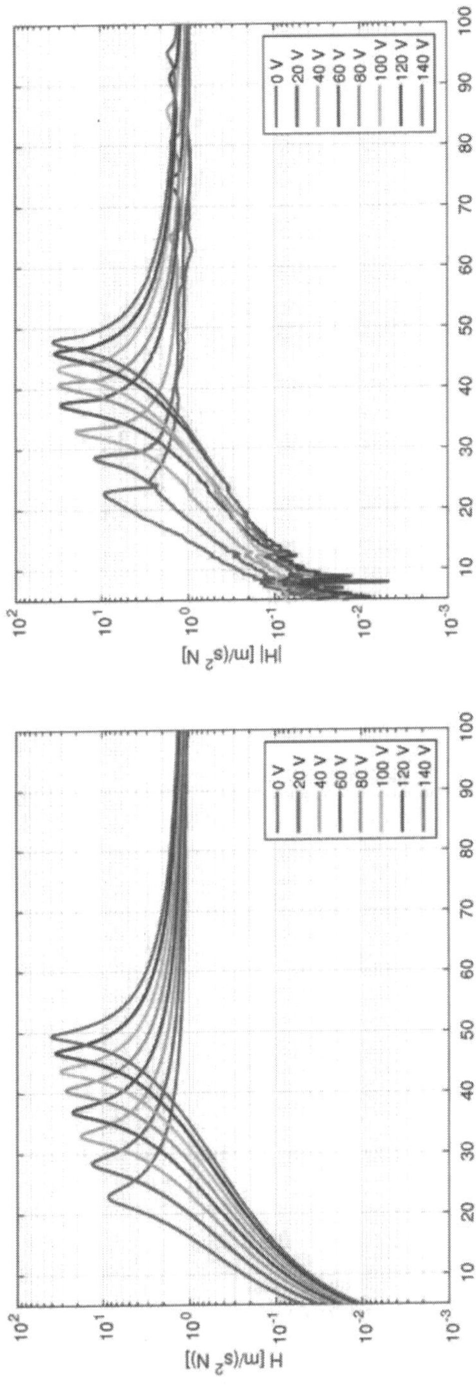

Figure 9. Frequency response functions of the adaptive vibration absorber for different DC voltages.

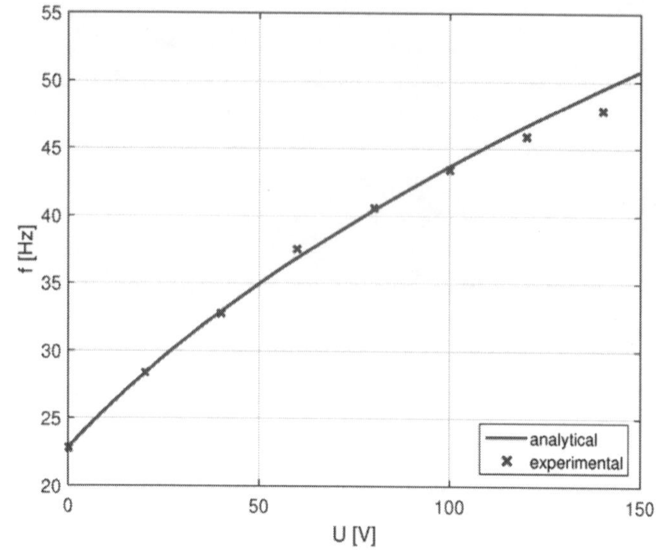

Figure 10. Comparison of resonance frequencies (analytical and experimental) for the electrical tuning of the adaptive vibration absorber.

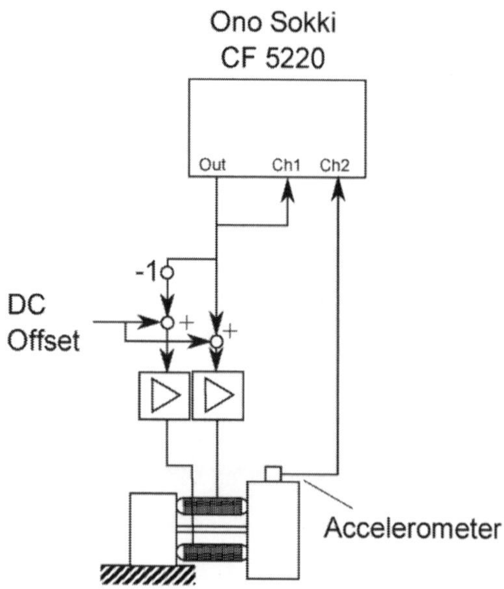

Figure 11. Test set up for the measurement of the dynamic block force of the actuator.

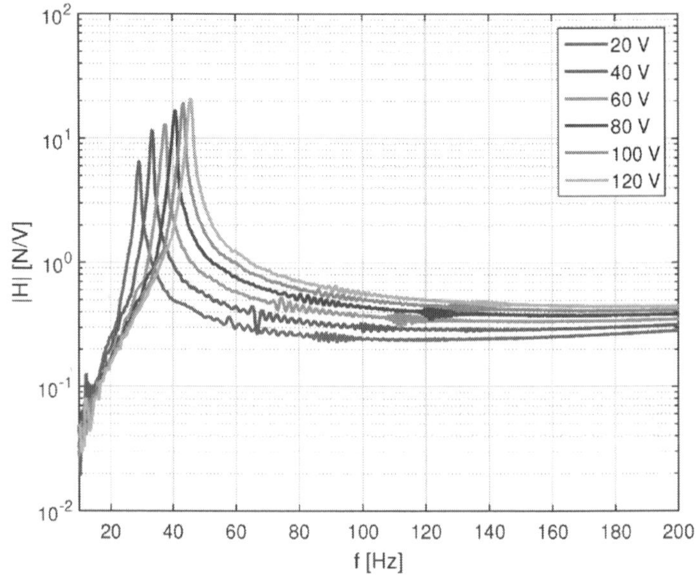

Figure 12. Measurement of the dynamic block force of the actuator.

5. APPLICATION TO AN ACTIVE VIBRATION CONTROL SYSTEM

To evaluate the performance in an active vibration control system, the adaptive absorber was mounted to a lightweight truss structure (Figure 13). It was instrumented with accelerometers at its base and at the inertial mass.

In the first step, the adaptive absorber was tuned by mechanical pre-load to a structural mode, which resulted in a vibration absorption effect at 43Hz43Hz (Figure 14, left). This effect was enhanced by choosing an additional electrical pre-load of 50VDC50VDC, which resulted in an absorption frequency of 50Hz50Hz. Although this frequency does not match the structural resonance at 48Hz48Hz exactly, this configuration served well as a basis for adjusting the active control system.

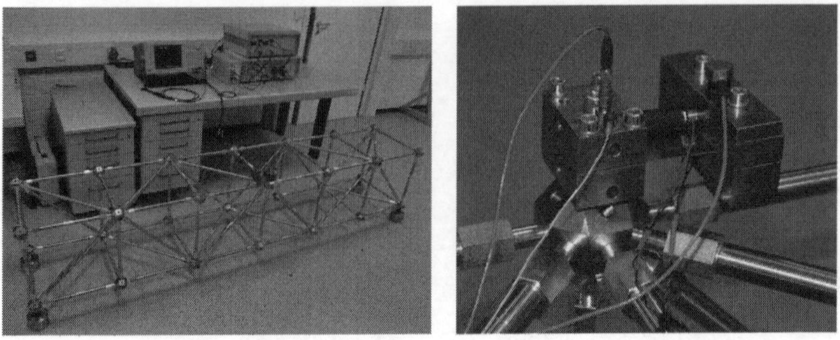

Figure 13. Test set up for the active control experiment (**left**) and detail of the actuator instrumentation (**right**).

In the next step, the active control system was set up and tuned. Two control loops were implemented successively and connected (Figure 15). First, an active velocity feedback loop Hv(s)Hv(s) was used to enhance the damping of the inertial mass actuator, which caused the better performance of the absorption at the first mode at 48 Hz and a higher robustness against the interaction between different control loops. In order to derive the dynamic velocity from the measured acceleration at the tip mass, an integrator is used. In a practical control system, this is implemented with a low pass filter:

$$H_v(s) = \frac{g_v}{s + \omega_{LP}} \tag{13}$$

The cut-off frequency ωLPωLP is set well below the resonance of the actuator, here at about 5 Hz, and gvgv is chosen appropriately.

Second, an acceleration feedback loop was tuned to the second mode of the truss at 125 Hz (Figure 14, right). The corresponding transfer function is a second order low pass filter:

$$H_a(s) = \frac{g_a \omega_c}{s^2 + 2\theta\omega_c s + \omega_c^2} \tag{14}$$

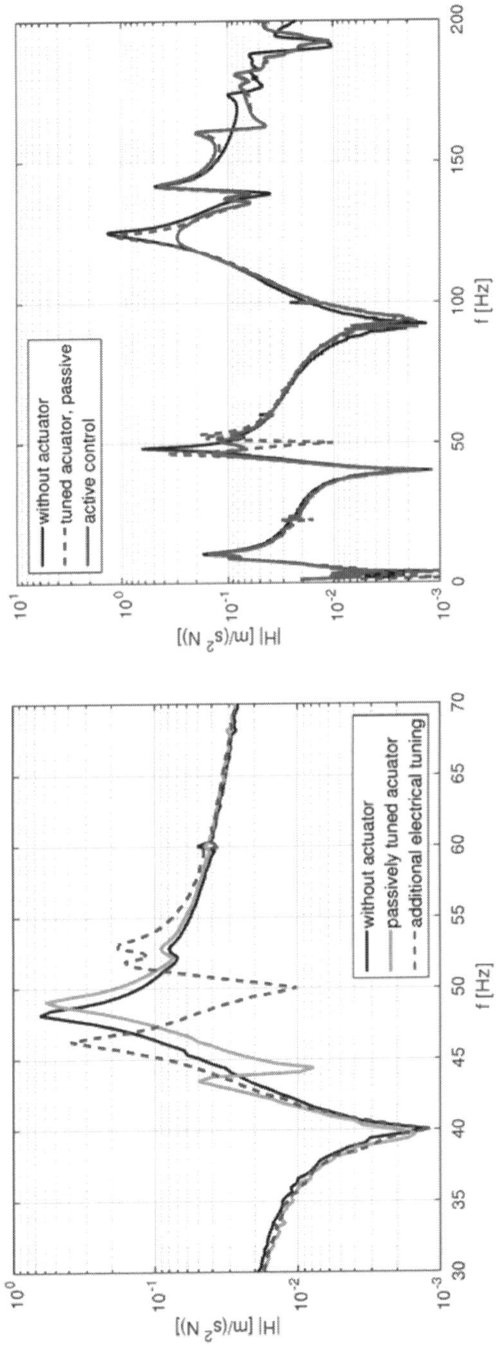

Figure 14. Tuning of the adaptive absorber to the first elastic mode of the truss and control results.

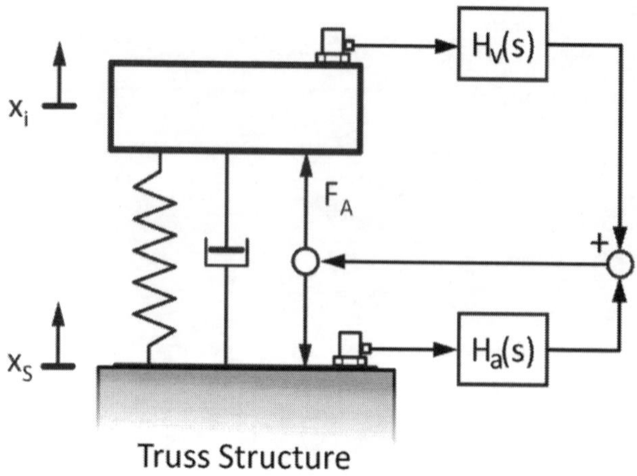

Figure 15. Block diagram of the control system.

Here, ωc is the tuning frequency and θ the damping coefficient, while gaga is the control loop gain. The control system was implemented with analogue circuitry.

Since the actuator is not symmetric, a bending moment is exciting at its base additional to the normal force. This results in lowering of a higher resonance frequency from 180 Hz to 160 Hz and a deterioration of the vibration amplitudes there.

6. RESULTS AND DISCUSSION

A piezoelectric adaptive vibration absorber was introduced that combines the possibility to tune the resonance frequency by static pre-stress forces with the option to generate dynamic signals, which can be used for tuning of the absorber's damping, but also for active vibration control in the frequency range above the resonance frequency. The properties were first evaluated by an analytical model and validated by experimental results. A good match between the predictions by the model and the measurements was observed.

Furthermore, a relatively wide tuning range for the absorber's resonance frequency was gained, which is comparable to other concepts using servo motors for tuning [5].

Furthermore, measurements of the generated dynamic force for the utilization as an inertial mass actuator were conducted. While generally, the block force was nearly constant above the resonance frequency, a slight dependency on the applied pre-loading voltage was observed. The vibration control experiments on a laboratory structure demonstrated the versatility of the concept. Tuning was possible using mechanically- and electrically-applied pre-loads. The damping could be adjusted by a velocity feedback loop, which is an interesting alternative to the commonly-used rubber elements in other absorber designs. Furthermore, the possibility for enhancement of the vibration control performance by using the device as an inertial mass actuator was successfully shown.

7. CONCLUSIONS

The results have shown that the design presented in this paper possesses the potential for vibration control applications. Generally, a wide tuning range of the adaptive absorber combined with the option to actively influence the absorber characteristics or to implement active control systems make this design an interesting alternative to other concepts.

However, there are also some existing design issues, which limit the performance. A certain level of pre-stress force should be applied in order to keep the properties of a single-degree-of-freedom oscillator, which limits the tuning range of the concept. Furthermore, the dual use as adaptive-passive and active device might be restricted: as the pre-loading voltage reaches the positive or negative maximum values of the used power amplifier, the maximum amplitude of the dynamic actuation should be decreased in order to avoid clipping effects. Furthermore, the dependency of the block force to the pre-load voltage should be considered when designing the active control loop; either adaptive control concepts like the filtered-X-LMS [17], parametric change of the gain with respect to the DC voltage or a design with a certain robustness to gain variations could be used.

Furthermore, further investigations and improvements of the adaptive absorber are necessary in order to enable the application outside the laboratory. The performance of the absorber in case of larger vibration amplitudes of the host structure should be analyzed; the reliability tests already performed with other types of adaptive absorbers should also be conducted with the design introduced in this paper. Finally, the mechanical design could be further improved towards a more compact solution.

ACKNOWLEDGMENTS

This work was partly conducted within the framework of the LOEWE center AdRIA. The support by the state of Hesse is gratefully acknowledged. Furthermore, the support by our colleague Tobias Röglin, who conducted the general design of the adaptive absorber, is kindly acknowledged.

AUTHOR CONTRIBUTIONS

D. Mayer conceived the vibration absorber. S. Herold derived the analytical models. D. Mayer and S. Herold performed the experiments, analyzed the data, and wrote the paper.

CONFLICTS OF INTEREST

The authors declare no conflict of interest.

REFERENCES

1. Frahm, H. Device for damping vibrations of bodies. US Patent 989,958, 18 April 1911.

2. Brennan, M.J. Some Recent Developments in Adaptive Tuned Vibration Absorbers/Neutralisers. *Shock Vib.* **2006**, *13*, 531–543.

3. Wright, R.I.; Kidner, M.R.F. Vibration Absorbers: A Review of Applications in Interior Noise Control of Propeller Aircraft. *J. Vib. Control* **2004**, *10*, 1221–1237.

4. Bonello, P. Adaptive Tuned Vibration Absorbers: Design Principles, Concepts and Physical Implementation. In *Vibration Analysis and Control - New Trends and Developments*; Beltran-Carbajal, F., Ed.; InTech: Rijeka, Croatia, 2011.

5. Mayer, D.; Pfeiffer, T.; Vrbata, J.; Melz, T. Adaptive-passive vibration control systems for industrial applications. *Proc. SPIE* **2015**, *9433*.

6. Winberg, M.; Johansson, S.; Claesson, I. Inertial mass actuators, understanding and tuning. In Proceedings of the ICSV 11, St. Petersburg, Russia, 5–8 July 2004.

7. Konstanzer, P.; Grünewald, M.; Jänker, P.; Storm, S. Piezo Tuneable Vibration Absorber System for Aircraft Interior Noise Reduction. In Proceedings of the Euronoise 2006, Tampere, Finland, 30 May–1 June 2006.

8. Herold, S.; Mayer, D.; Melz, T.; Röglin, T. Design and Test of a Piezoelectric Inertial Mass Actuator for Active Vibration Control.

In *Vibration Engineering and Technology of Machinery*; Sinha, J., Ed.; Mechanisms and Machine Science, Springer International Publishing: Cham, Switzerland, 2014; Volume 23, pp. 587–597.

9. Mayer, D.; Kauba, M.; Herold, S.; Koch, T. Approaches for distributed active and passive vibration compensation. In Proceedings of the ISMA 2010, 24th International Conference on Noise and Vibration Engineering, Leuven, Belgium, 20–22 September 2010; p. 13.

10. Hagood, N.W.; Crawley, E.F. Experimental investigation into passive damping enhancement for space structures. *J. Guidance Control Dyn.* **1991**, *14*, 1100–1109.

11. Preumont, A. *Vibration Control of Active Structures*; Springer: Dordrecht, Netherlands, 2011.

12. Crawley, E.F.; Luis, J.D. Use of piezoelectric actuators as elements of intelligent structures. *AIAA J.* **1987**, *25*, 1373–1385.

13. Benjeddou, A. Advances in piezoelectric finite element modeling of adaptive structural elements: A survey. *Comput. Struct.* **2000**, *76*, 347–363.

14. Ruschmeyer, K. *Piezokeramik: Grundlagen, Werkstoffe, Applikationen*; Expert-Verl.: Renningen-Malmsheim, Germany, 1995.

15. Herold, S.; Mayer, D.; Hanselka, H. Transient Simulation of Adaptive Structures. *J. Intell. Mater. Syst. Struct.* **2004**, *15*, 215–224.

16. Herold, S.; Atzrodt, H.; Mayer, D.; Thomaier, M. Modeling approaches for active systems. *Proc. SPIE* **2006**, *6173*.

17. Elliott, S. *Signal Processing for Active Control*; Academic Press: Cambridge, MA, USA, 2000.

CHAPTER 10

Feedback control unit with an inerter proof-mass electrodynamic actuator

Michele Zilletti

ABSTRACT

In this study the use of an inerter is considered for active vibration control of a structure excited by white noise. The structure is modelled as a single degree of freedom system and the control system consists of a vibration absorber with a mass suspended on a spring, a damper and an inerter. The absorber is equipped with a reactive force transducer in parallel with the passive suspension which is driven with a signal proportional to the velocity of the structure under control measured by an ideal collocated sensor. The effect of the inerter on the control stability and performance of the control system is investigated. It is shown that the effect of the inerter is to reduce the natural frequency of the inertial actuator, improving the stability of the feedback loop and thus its performance. The optimisation of the physical and control parameters of the control system such as the internal damping of the actuator, its natural frequency, its inertance and the feedback gain are considered such that either the kinetic energy of the host structure is minimised or the power dissipated by the control system is maximised.

KEYWORDS

Velocity feedback; Inerter; Inertial actuator; Proof-mass actuator

1. INTRODUCTION

Passive tuned vibration absorbers were proposed by Farhm [1] in 1911 and have been widely used to control structural vibrations ever since. They normally consist of a mass suspended on an elastic mount, which can usually be modelled as a single degree of freedom system. When they are used to control the vibration of flexible structures subjected to broadband excitation their natural frequency and internal damping are tuned to reduce the structural response in a narrow frequency band around a structural resonance [2]. Many optimisation criteria of passive vibration absorbers have been proposed depending on the application and a vast literature can be found on this topic [2],[3], [4], [5], [6], [7], [8], [9], [10] and [11]. As summarised by Zilletti et al. [11] several criteria have been derived using lumped parameter models. One drawback of passive vibration absorbers is that they can only be tuned to one resonance frequency of the host structure at the time and very small reduction is achieved at other resonance frequencies. Therefore multi-degrees of freedom absorbers [12] or time variant one [13],[14] and [15] have been proposed in order to widen the frequency band of operation of these devices. However the efficiency of passive and semi-active vibration absorbers increases with the mass ratio of the absorber to the system under control resulting in high added weight.

A considerable improvement can be achieved by using active vibration absorbers, which consist of a passive vibration absorber with a reactive force transducer in parallel with the suspension, which is driven to implement a feedback loop. An example of this device is the voice coil inertial actuator with an accelerometer sensor at its footprint commonly used in vibration control applications [16]. The advantages of an active vibration absorber are its adaptability to parameters changes of the structure under control, which guarantees a better vibration control performance in a wider range of operating conditions compared with a passive one. Depending on the circumstances, active solutions may be cheaper and lighter than passive systems offering performances that no passive system can achieve, especially in controlling low frequency vibration. One way of driving these devices in order to damp the structural response of flexible structures subjected to broadband excitation is by using a velocity feedback loop [17],[18], [19] and [20]. In this case the controller acts as a sky-hook active damper in the frequency region above the resonance of the actuator, before higher order resonances interfere with their dynamics. This type of control system is only conditionally stable because of the 180° phase shift in the response of the actuator due to its resonance. In order to improve the stability of the controller and thus its performance, the natural frequency of the absorber has to be as low as possible compared to the natural frequency of the structure under control [21]. This may be achieved by using a low suspension stiffness which, however, may cause problems with the static deflection of the suspended mass due to gravity. This can be overcome with internal displacement feedback loops [22]. In a similar manner, internal velocity feedback loops can be used to increase the internal damping [23]. These methods require external power and particular care has to be taken in the practical implementation of the feedback loop to avoid instability phenomena. Another strategy to improve the stability of these devices is by using a compensator which electronically shifts the resonance frequency of the device at lower frequency with the drawback of increasing the feedback signal in the low frequency region which may result in a stroke/force saturation effect of the actuator [24].

In this paper the use of an inerter that lowers the fundamental resonance frequency of the actuator is investigated. The inerter is a mechanical device which produces a force proportional to the relative acceleration of its two terminals and was initially proposed by Smith [25]. The constant of proportionality is called inertance and is measured in kilogram. Since its invention inerters have been widely used and a vast literature can be found on their applications. Although a complete literature review on this topic is beyond the scope of this paper, it is instructive to consider few examples of inerters in vibration control applications. Inerters have been previously used in vibration absorbers with the aim of improving their performance in damping out the vibration of both linear [26] and nonlinear systems [27]. Also, vibration absorbers with tuneable inertance have been proposed which use a continuously variable transmission and gear ratio control system[28]. Inerters have been used to improve the performance of passive [29], [30] and [31]and semi-active car suspensions [32]. In the case of semi-active suspensions the inertance is electronically synthetised with a state- feedback control system. Several applications of inerters can be found in different systems such as passive network synthesis [33], [34] and [35], motorcycles steering system [36] and [37], train suspensions [38] and [39], civil engineering applications [40], and landing gear suspensions [41] with the aim of improving vibration control performance.

Chen et al. [42] have demonstrated that the natural frequency of a lumped parameter mechanical system with an inerter can always be reduced by increasing the inertance. This suggests that the use of an inerter in an active vibration absorber could be used to improve its control stability by lowering the resonance frequency of the device without softening the suspension.

In this paper the proposed vibration absorber that implements an ideal velocity feedback loop is used to control a single mode of a vibrating structure, modelled as a single degree of freedom system, which is subjected to white noise excitation. Throughout the paper, to better emphasise the advantages of the proposed control unit, the vibration control effect of the proposed device with the inerter is compared with the effect produced by a reference control unit without the inerter. It is demonstrated that the use of the inerter improves the control stability and thus provides a better control performance compare to the configuration without the inerter.

One important issue with such a control system is how the feedback gain is set to optimally control the vibration of the hosting structure. In an ideal system, the optimal feedback gain would be adjusted to minimise the kinetic energy of the structure under control. However this optimisation may be difficult to implement in practice because it would require velocity measurements in many points of the structure. Previous studies have shown that the maximisation of power absorbed by the controller, which can be easily estimated from the feedback signal, is equivalent to the minimisation of the kinetic energy of the structure under control subjected to a white noise excitation [43],[44] and [45]. This paper extends this result to the case of the control unit with the inerter.

The organisation of the paper is as follow. In Section 2 the mathematical model of the primary system with the controller is derived providing the expressions for the frequency response functions. In Section 3 the stability of the control system is analytically assessed and the expressions for the kinetic energy of the system under control and the power absorbed by the controller are derived. In Section 4 numerical simulation results are presented and conclusions are drawn in Section 5.

2. MATHEMATICAL MODEL

Typically when an inertial actuator is used to implement a velocity feedback loop to control the resonant response due to one mode of a flexible structure, a simplified lumped parameters model is considered. As shown in Fig. 1 it is assumed that the primary structure is composed by a modal mass m_1, a damper c_1 and a stiffness k_1 and is excited by a primary force f_p.

Classical inertial actuators are composed by a block mass m_2, mounted on a suspension of stiffness k_2 a damping c_2. In this study an inerter of inertance b_2, producing a force proportional to the relative acceleration between mass m_1 and m_2 has been added in parallel with a reactive force transducer. At frequencies above its fundamental resonance frequency, the inertial actuator produces a sky-hook active forcef_a proportional to the absolute velocity of the modal mass m_1 via the control gain g.

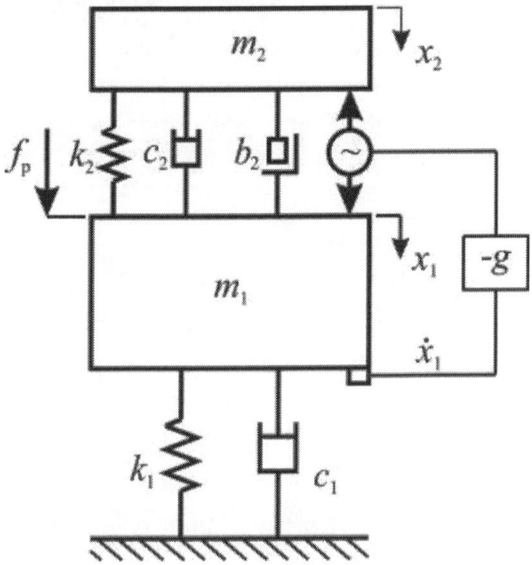

Figure. 1. Scheme of the SDOF system with a velocity feedback loop using an inertial actuator.

The equation of motion of the system shown in Fig. 1 can be written in a matrix form as:

$$\mathbf{M}\ddot{\mathbf{x}}(t)+\mathbf{C}\dot{\mathbf{x}}(t)+\mathbf{K}\mathbf{x}(t)=\boldsymbol{f}(t) \quad (1)$$

where **M** is the mass matrix, **K** is the stiffness matrix and **C** is the damping matrix given by:

$$\mathbf{M}=\begin{bmatrix} m_1+b_2 & -b_2 \\ -b_2 & m_2+b_2 \end{bmatrix}, \mathbf{K}=\begin{bmatrix} k_1+k_2 & -k_2 \\ -k_2 & k_2 \end{bmatrix}, \mathbf{C}=\begin{bmatrix} c_1+c_2+g & -c_2 \\ -c_2-g & c_2 \end{bmatrix}, \quad (2)$$

$\mathbf{x}(t)=[x_1(t)\ \ x_2(t)]^T$ is the column vector containing the displacements of the two massesx_1 and x_1 and $\mathbf{f}(t)=\begin{bmatrix} f_p(t) & 0 \end{bmatrix}^T$ is the column vector of the primary excitation. It is important to notice that due to the feedback loop the damping matrix **C** is non-symmetric. As it will be shown in the next section this implies that the control system is only conditionally stable.

Assuming the excitation to be harmonic for the time being and expressing the force and the steady-state response in exponential form, Eq. (1) becomes:

$$\mathbf{S}(j\omega)\mathbf{x}(j\omega)=\mathbf{f}(j\omega) \tag{3}$$

where

$$\mathbf{S}(j\omega)=-\omega^2\mathbf{M}+j\omega\mathbf{C}+\mathbf{K} \tag{4}$$

is the dynamic stiffness matrix. The solution of Eq. (3) can be obtained as:

$$\mathbf{x}(j\omega)=\mathbf{S}^{-1}(j\omega)\mathbf{f}(j\omega). \tag{5}$$

Integrating Eq. (5) to obtain the velocities yields:

$$\dot{\mathbf{x}}(j\omega)=\mathbf{Y}(j\omega)\mathbf{f}(j\omega), \tag{6}$$

where $\dot{\mathbf{x}}(j\omega)=j\omega\mathbf{x}(j\omega)$ and $\mathbf{Y}(j\omega)=j\omega\mathbf{S}^{-1}(j\omega)$ is the mobility matrix. Using the expression of **M**, **K** and **C** of Eq. (2), the steady state complex response of the system can be expressed in terms of the input and transfer frequency response functions (FRF) as:

$$Y_{11}(j\omega)=\frac{\dot{x}_1}{f_p}=$$

$$=\frac{jk_2\omega-c_2\omega^2-jm_2\omega^3-jb_2\omega^3}{k_1k_2+jc_2k_1\omega+jc_1k_2\omega-c_1c_2\omega^2-b_2k_1\omega^2-k_2m_1\omega^2-k_1m_2\omega^2-k_2m_2\omega^2-jb_2c_1-jc_2m_1\omega^3-}$$

$$jc_1m_2\omega^3-jc_2m_2\omega^3+gm_2\omega^2+b_2m_1\omega^4+b_2m_2\omega^4+m_1m_2\omega^4$$

(7)

$$Y_{21}(\mathrm{j}\omega)=\frac{\dot{x}_2}{f_p}=$$

$$=\frac{\mathrm{j}k_{2}\omega-c_{2}\omega^2+g\omega^2-\mathrm{j}b_{2}\omega^3}{k_1k_2+\mathrm{j}c_2k_{1}\omega+\mathrm{j}c_1k_{2}\omega-c_1c_{2}\omega^2-b_2k_{1}\omega^2-k_2m_{1}\omega^2-k_1m_{2}\omega^2-k_2m_{2}\omega^2-\mathrm{j}b_2c_1-\mathrm{j}c_2m_{1}\omega^3-}$$

$$\mathrm{j}c_1m_{2}\omega^3-\mathrm{j}c_2m_{2}\omega^3+gm_{2}\omega^2+b_2m_{1}\omega^4+b_2m_{2}\omega^4+m_1m_{2}\omega^4$$

(8)

where $\dot{x}_1$ and $\dot{x}_2$ are the complex frequency dependent velocities of masses m_1 andm_2 respectively. The two FRFs can be expressed in non-dimensional form as follows:

$$\Gamma(\mathrm{j}\lambda)=\frac{B_0+(\mathrm{j}\lambda)B_1+(\mathrm{j}\lambda)^2B_2+(\mathrm{j}\lambda)^3B_3}{A_0+(\mathrm{j}\lambda)A_1+(\mathrm{j}\lambda)^2A_2+(\mathrm{j}\lambda)^3A_3+(\mathrm{j}\lambda)^4A_4}$$ (9)

$$\Theta(\mathrm{j}\omega)=\frac{C_0+(\mathrm{j}\lambda)C_1+(\mathrm{j}\lambda)^2C_2+(\mathrm{j}\lambda)^3C_3}{A_0+(\mathrm{j}\lambda)A_1+(\mathrm{j}\lambda)^2A_2+(\mathrm{j}\lambda)^3A_3+(\mathrm{j}\lambda)^4A_4},$$ (10)

where the coefficients A_{0-4}, B_{0-3} and C_{0-3} are given by:

$$\begin{array}{lll}
A_0=\nu^2 & B_0=0 & C_0=0\\
A_1=2\zeta_2\nu+2\zeta_1\nu^2 & B_1=\nu^2 & C_1=\nu^2\\
A_2=\nu^2+1+\mu\nu^2+4\zeta_1\zeta_2\nu+\delta & B_2=2\zeta_2\nu & C_2=2\zeta_2\nu+4\alpha\zeta_2\nu\\
A_3=2\zeta_2\nu+2\zeta_2\mu\nu+2\zeta_1+4\alpha\nu\zeta_2\mu+2\zeta_1\delta & B_3=1+\delta & C_3=\delta\\
A_4=1+\delta+\delta\mu & &
\end{array}$$

The seven non-dimensional parameters in Eqs. (9) and (10) are defined by:

$$\mu=m_2/m_1\,\nu=\omega_2/\omega_1\,\lambda=\omega/\omega_1\,\zeta_1=c_1/(2m_1\omega_1)\,\zeta_2=c_2/(2m_2\omega_2)\,\alpha=g/(2c_2)\,\delta=b_2/m_2$$

(11)

where μ is the mass ratio, ν is the frequency ratio, λ is the normalised driving frequency, ζ_1 and ζ_2 are the primary and secondary damping ratios respectively, α is the normalised velocity control feedback gain and δ is the normalised inertance. The natural frequency of the primary system, ω_1, and the natural frequency of the inertial actuator, ω_2, are defined as:

$$\omega_1 = \sqrt{k_1/m_1},$$
$$\omega_2 = \sqrt{k_2/m_2}. \qquad (12)$$

3. STABILITY AND PERFORMANCE

In this section the stability of the control system is analytically assessed and the expressions for the kinetic energy of the primary system and the power absorbed by the controller are derived.

3.1. Stability of the controller

It is important to notice that due to the resonance of the inertial actuator the velocity loop is only conditionally stable thus the control gain cannot exceed the value that leads the system to instability. Fig. 2 shows the block diagram of the control system where $\Delta(j\lambda)$ is the open loop frequency response function given by:

$$\Delta(j\lambda)=\Theta_0(j\lambda)-\Gamma_0(j\lambda). \qquad (13)$$

where $\Gamma_0(j\lambda)$ and $\Theta_0(j\lambda)$ are obtained from Eqs. (9) and (10) by setting α to zero.

The stability of the controller can be assessed from the open loop frequency response function applying the Routh's stability criterion. The characteristic equation of the control system is given by the denominator of Eq. (13) and the related Routh's coefficients are listed in Table 1.

The coefficients of the characteristic equation are all positive and so is the coefficient h_1. To guarantee the stability of the system the numerator of coefficient h_2 has to be greater than zero and thus the following condition has to be satisfied:

$$S_2\alpha^2+S_1\alpha+S_0>0 \tag{14}$$

where S_2, S_1 and S_0 are given by:

$$\begin{gathered} S_2=2\nu(\mu\nu^3\zeta_1^2+\zeta_1\Big((1+\delta)^2-2(1+\delta+\delta\mu)\nu^2+(1+\mu)^2\nu^4+4(1+\delta)\nu^2\zeta_1^2\Big)\zeta_2+\nu(\mu+4(1+\delta \\ +(1+\mu)\nu^2)\zeta_1^2)\zeta_2^2+4(1+\mu)\nu^2\zeta_1\zeta_2^3) \\ S_1=4\mu\nu^2\zeta_2((1+\delta-(1+\mu)\nu^2)\zeta_2+4\nu^2\zeta_1^2\zeta_2+\nu\zeta_1(-1-\delta+(1+\mu)\nu^2+4\zeta_2^2)) \\ S_0=-8\mu^2\nu^4\zeta_2^2 \end{gathered} \tag{15}$$

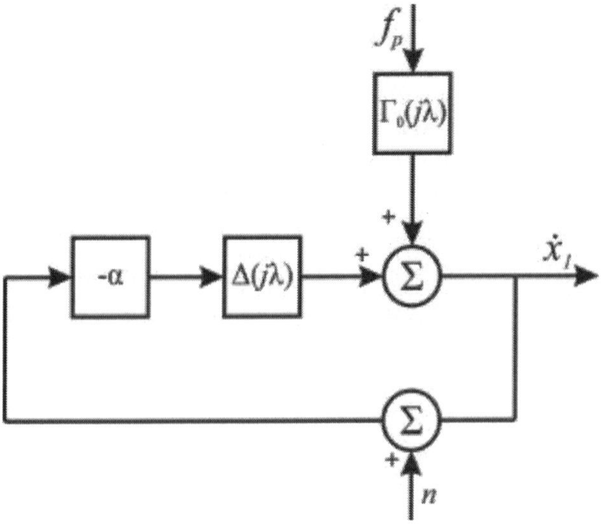

Figure. 2. Block diagram of the control system with measurement noise n.

Table 1. Routh's coefficients chart.

A_4	A_2	A_0
A_3	A_1	
$h_1=\frac{A_3A_2-A_4A_1}{A_3}$	A_0	
$h_2=\frac{h_1A_1-A_3A_0}{h_1}$		
A_0		

The block diagram shown in Fig. 2 also includes the signal n which represents measurement errors or noise on the feedback signal. The block diagram reduction indicates that the closed-loop output is given by:

$$\dot{x}_1 = \frac{\Gamma_0(j\lambda)}{1+g\Delta(j\lambda)} f_p - \frac{\Gamma_0(j\lambda)\alpha\Delta(j\lambda)}{1+\alpha\Delta(j\lambda)} n = S(j\lambda) f_p - T(j\lambda) n \tag{16}$$

where the transfer functions S(jλ) and T(jλ) are usually defined as sensitivity and complementary sensitivity function respectively. Fig. 3 shows the magnitude of S(jλ)(solid line) and T(jλ) (dashed line) functions for low values of control gain (plot (a)) and for the control gain approaching the maximum stable gain (plot (b)).

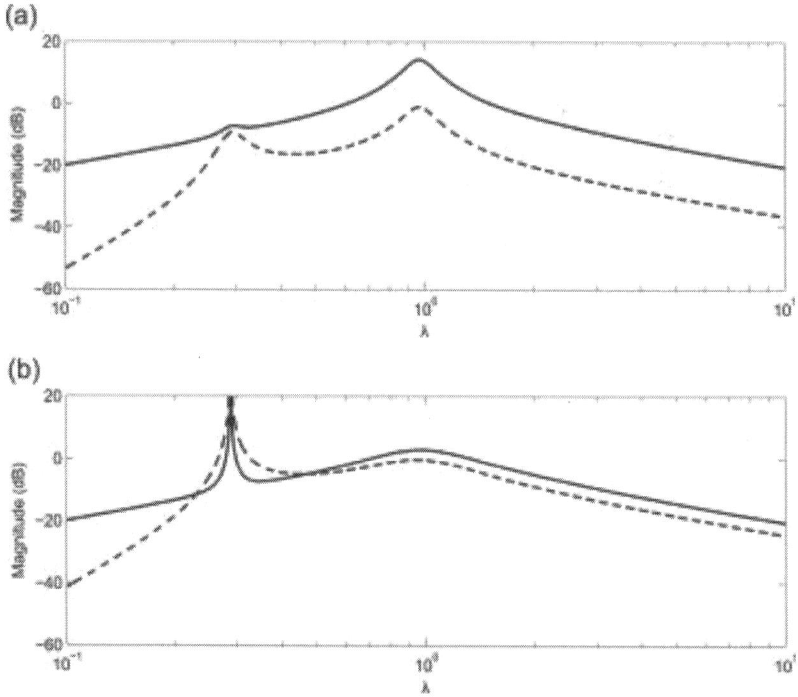

Figure 3. Magnitude of the sensitivity (solid line) and the complementary sensitivity (dashed line) functions for $\alpha_{max}/4$ (plot (a)) and for the control gain approaching the maximum stable gain α_{max} (plot (b)).

Plot (a) shows that the magnitude of T(jλ) is lower than S(jλ) thus the control system has a good feedback noise rejection. However as shown in plot (b), when the control gain approaches the maximum stable gain the magnitude of T(jλ) gets larger than S(jλ) in the frequency region around the natural frequency of the control unit. This brief analysis shows that particular care must be taken in reducing the noise in the feedback signal since an excessive noise level may limit the control gain that can be achieved in practice and thus degrade the performance of the system.

3.2. Energy and power formulation

The performance of the controller discussed in the next section is assessed in terms of the reduction of the kinetic energy of the system under control when excited by a broadband white noise force. The kinetic energy of the primary mass per unit excitation force with a uniform power spectral density (PSD) $S_f(\omega)$ measured in N^2 s/rad is given by:

$$E_k = \frac{m_1 E\left[|\dot{x}_1(t)|^2\right]}{2S_f\omega_1/k_1}, \tag{17}$$

where $E[\,]$ denotes the expectation value and the constant $S_f\omega_1/k_1$ is introduced to ensure that E_k is dimensionless. The mean squared value of the velocity of the primary mass, m_1, can be written as:

$$E[|\dot{x}_1(t)|^2] = \frac{S_f\omega_1}{m_1 k_1}\int_{-\infty}^{+\infty} |\Gamma(j\lambda)|^2 d\lambda. \tag{18}$$

Thus substituting Eq. (9) in (18) and then in (17) yields:

$$E_k = \frac{1}{2}\int_{-\infty}^{+\infty} \left| \frac{B_0 + (j\lambda)B_1 + (j\lambda)^2 B_2 + (j\lambda)^3 B_3}{A_0 + (j\lambda)A_1 + (j\lambda)^2 A_2 + (j\lambda)^3 A_3 + (j\lambda)^4 A_4} \right|^2 d\lambda. \tag{19}$$

Eq. (19) can be integrated using the formula in reference [46] leading to:

$$E_k = \frac{1}{2} \frac{G_1 \alpha + G_0}{F_2 \alpha^2 + F_1 \alpha + F_0} \tag{20}$$

where the coefficients $G_{0,1}$ and $F_{0,1,2}$ are given in Table A.1 in Appendix A. The optimal control gain αopt that minimise the kinetic energy of the primary system can be calculated by setting to zero the derivative of E_k with respect to α leading to:

$$F_2 G_1 \alpha^2 + 2 F_2 G_0 \alpha + F_1 G_0 - F_0 G_1 = 0. \tag{21}$$

As shown in references [43], [44] and [47] a convenient way to tune the control gain is by maximise the power absorbed by the control unit. The total power absorbed by the control unit is given by the sum of the passive power dissipated by the damper and the active power of the electromechanical transducer. The PSD of the power absorbed by the passive damper c_2 is equal to:

$$P_{pas}(\omega) = \frac{1}{2} \mathrm{Re}\{f_d^*(j\omega)[\dot{x}_1(j\omega) - \dot{x}_2(j\omega)]\}, \tag{22}$$

where * denotes complex conjugate and the force f_d is the force produced by the damper given by:

$$f_d(j\omega) = c_2(\dot{x}_1(j\omega) - \dot{x}_2(j\omega)). \tag{23}$$

Substituting Eq. (23) in (22) the power passively dissipated by the inertial actuator becomes:

$$P_{pas}(\omega) = \frac{1}{2} c_2 |\dot{x}_1(j\omega) - \dot{x}_2(j\omega)|^2, \tag{24}$$

The non-dimensional power dissipated by the damper per unit excitation force is given by:

$$I_{pas} = \frac{c_2 E\left[|\dot{x}_1(t) - \dot{x}_2(t)|^2\right]}{4\pi S_f \omega_1 / k_1}, \tag{25}$$

which represents the ratio of the power passively dissipated by the control unit to that generated by the excitation force with spectrum density S_f acting on a damper of value$k_1/(\pi\omega_1)$. The mean squared value of the relative velocity is given by:

$$E[|\dot{x}_1(t)-\dot{x}_2(t)|^2]=\frac{S_f\omega_1^2}{k_1}2\zeta_2\mu\nu\int_{-\infty}^{+\infty}|\Gamma(j\lambda)-\Theta(j\lambda)|^2 d\omega. \tag{26}$$

Substituting Eqs. (9) and (10) in Eq. (26) and then in (25) the power passively dissipated by the damper becomes:

$$I_{pas}=\frac{\zeta_2\mu\nu}{2\pi}\int_{-\infty}^{+\infty}\left|\frac{D_0+(j\lambda)D_1+(j\lambda)^2D_2+(j\lambda)^3D_3}{A_0+(j\lambda)A_1+(j\lambda)^2A_2+(j\lambda)^3A_3+(j\lambda)^4A_4}\right|^2 d\omega, \tag{27}$$

where

$$D_0=C_0-B_0=0 D_1=C_1-B_1=0 D_2=C_2-B_2=4\alpha\nu\zeta_2 D_3=C_3-B_3=-1 \tag{28}$$

The integral over the frequency band between $\pm\infty$ in Eq. (27) can be calculated using the expression given in reference [46], leading to:

$$I_{pas}=\frac{1}{2}\frac{H_2\alpha^2+H_1\alpha+H_0}{F_2\alpha^2+F_1\alpha+F_0}, \tag{29}$$

where the coefficients $H_{0,1,2}$ are given in Table A.1 in Appendix A.

The PSD of the active power produced by the electromechanical reactive force is equal to:

$$P_{act}(\omega)=\frac{1}{2}\mathrm{Re}\{f_a^*(j\omega)[x^1(j\omega)-x^2(j\omega)]\}. \tag{30}$$

Assuming that the force f_a is proportional to the velocity $\dot{x}_1(j\omega)$ of the primary mass, Eq.(30) can be written as:

$$P_{act}(\omega)=-\frac{g}{2}\mathrm{Re}\{\dot{x}_1^*(j\lambda)[\dot{x}_1(j\omega)-\dot{x}_2(j\omega)]\}=-\frac{g}{2}[|\dot{x}_1(j\omega)|^2-\mathrm{Re}\{\dot{x}_1^*(j\omega)\dot{x}_2(j\omega)\}]. \tag{31}$$

After some manipulations Eq. (31) can be written as:

$$P_{act}(\omega) = -\frac{g}{4}\left[|\dot{x}_1(j\omega)|^2 + |\dot{x}_1(j\omega) - \dot{x}_2(j\omega)|^2 - |\dot{x}_2(j\omega)|^2\right]. \quad (32)$$

In this case the active power per unit excitation force is given by:

$$I_{act} = -\frac{gE[|\dot{x}_1(t)|^2 + |\dot{x}_1(t) - \dot{x}_2(t)|^2 - |\dot{x}_2(t)|^2]}{4\pi S_f \omega_1 / k_1}, \quad (33)$$

which represents the ratio of the power actively dissipated by the control unit to that generated by the excitation force with spectrum density S_f acting on a damper of value$k_1/(\pi\omega_1)$. The mean squared value of the relative velocity times the control gain gcan be expressed as follow:

$$gE\left[|\dot{x}_1(t)|^2 + |\dot{x}_1(t) - \dot{x}_2(t)|^2 - |\dot{x}_2(t)|^2\right] = \frac{S_f\omega_1^2}{k_1} 2\zeta_2 \mu\nu \left[\int_{-\infty}^{+\infty} |\Gamma|^2 d\lambda + \int_{-\infty}^{+\infty} |\Gamma - \Theta|^2 d\lambda - \int_{-\infty}^{+\infty} |\Theta|^2 d\lambda\right] \quad (34)$$

Substituting the expressions of Γ and Θ in Eq. (34) and calculating the integrals over the frequency band between $\pm\infty$ using the expression given in reference [46] the total active power becomes:

$$I_{act} = \frac{L_2\alpha^2 + L_1\alpha}{F_2\alpha^2 + F_1\alpha + F_0}, \quad (35)$$

where the coefficients $L_{1,2}$ are given in Table A.1 in Appendix A. The total power absorbed by the control unit, given by the sum of power dissipated by the passive damper I_{pas} and the active power produced by the electromechanical actuator I_{act}can be written as:

$$I_p = \frac{1}{2}\frac{(2L_2 + H_2)\alpha^2 + (2L_1 + H_1)\alpha + H_0}{F_2\alpha^2 + F_1\alpha + F_0}. \quad (36)$$

From the energy balance, in steady state conditions, the total power input into the whole system I_{pin} is equal to the power dissipated by damper 1 plus the total power dissipated by the control unit I_p. The power dissipated by damper 1 is given by:

$$P_1(\omega)=\frac{1}{2}c_1|\dot{x}_1(j\omega)|^2. \tag{37}$$

In order to obtain a dimensionless formulation, a dimensionless index relative to $P_1(\omega)$can be defined by:

$$I_{p1}=\frac{c_1 E[|\dot{x}_1(t)|^2]}{4\pi S_f\omega_1^2/k_1}=\frac{1}{2}\frac{c_1}{m_1}E_k. \tag{38}$$

The power input into the system per unit input force can be calculated as the sum of Eqs.(36) and (38) giving:

$$I_{pin}=I_{p1}+I_p=\frac{1+\delta}{2+2\delta+2\delta\mu}. \tag{39}$$

Eq. (39) indicates that, when the system is excited by a random white noise force, the power input into the system only depends on the masses, m_1 and m_2 and the inertanceb_2. Multiplying Eq. (39) by the constant $2\pi S_f\omega_1^2/k_1$ and setting the inertance δ to zero, the power input reduces to $1/(2m_1)$ thus it only depends on the mass of the primary system as shown in references [11] and [48].

Taking the derivative with respect to α of Eq. (39) yields:

$$\frac{\partial I_{p1}}{\partial\alpha}+\frac{\partial I_p}{\partial\alpha}=0 \tag{40}$$

and substituting Eq. (38) in (40) gives:

$$\frac{\partial I_p}{\partial\alpha}=-\frac{1}{2}\frac{c_1}{m_1}\frac{\partial E_k}{\partial\alpha}. \tag{41}$$

Eq. (41) indicates that, for given c_1 and m_1, the derivative with respect to α of I_p is proportional to the derivative of E_k thus the maximisation of the power absorbed by the control unit is equivalent to the minimisation of the kinetic energy of the structure under control.

4. NUMERICAL SIMULATIONS

This section discusses the stability and performance of the velocity feedback loop implemented to control the resonant response of a structure modelled as single degree-of-freedom lumped parameter model.

To better understand the dynamics of the system and compare the vibration control effect obtained with the control unit with and without the inerter numerical simulations have been carried out using the parameters listed in Table 2.

Table 2. Non-dimensional parameters used in the simulations.

Parameter	Value
Mass ratio μ	0.1
Natural frequency ratio ν	0.5
Damping ratio of the primary system ζ_1	0.01
Inertance δ	2
Damping ratio of the inertial actuator ζ_2	0.1

4.1. Stability

The stability of the control system is first assessed with the Nyquist criterion. Fig. 4 shows the Bode (a) and the Nyquist (b) diagrams of the open loop frequency response function (OL-FRF) given by Eq. (13). The dashed line shows the OL-FRF for the classical inertial actuator without the inerter while the solid line shows the OL-FRF of the inertial actuator with the inerter proposed in this study. In order to obtain a non-dimensional formulation the damping ratio of the control unit ζ_2 was defined in Eq. (11) to be independent from the inertance b. However the actual damping ratio of the control unit is indeed influenced by the presence of the inerter thus the actual damping ratio of the inertial actuator ζ_2' is defined as:

$$\zeta_2' = \frac{c_2}{2\sqrt{k_2(m_2+b_2)}}, \tag{42}$$

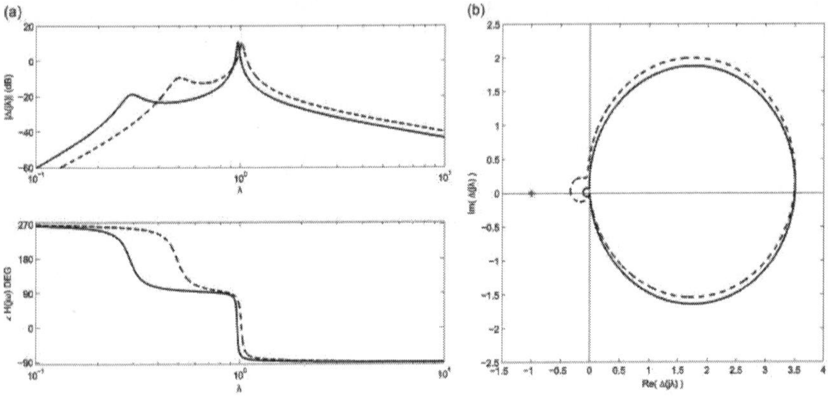

Figure. 4. (a) Bode and (b) Nyquist diagrams of the open loop FRF given in Eq. (13) when δ=0 (dashed line) andδ=2 (solid line) for the same value of $\zeta_2'=0.1$ and normalised to have the same maximum magnitude.

In order to compare the two configurations the value of the actual damping ratio ζ_2' of the two devices with and without the inerter has been set to 0.1. The OL-FRF in Fig. 4 has been normalised with respect to the magnitude at the second resonance peak in the OL-FRF of the classical configuration without the inerter, so that the differences between the two FRFs can be better compared.

Considering first the classical configuration without the inerter (dashed line) the magnitude of the OL-FRF shows two peaks at about the actuator fundamental resonance frequency (λ=0.5) and the fundamental resonance frequency of the system under control (λ=1). The two resonant responses are out of phase, thus the low frequency portion of the OL-FRF locus shown in Fig. 4(b) is characterised by a circle in the real-negative quadrants (first resonance) and a circle of larger size in the real-positive quadrants (second resonance). This indicates that the closed feedback loop is only conditionally stable with relatively low gain margin. Considering the case of the control unit with the inerter (solid line) the first peak of the OL-FRF is shifted at about λ=0.3 due to the effect of the inerter. The solid line in Fig. 4(b) shows that the circle of the first resonance in the real-negative quadrants is about three times smaller than the one of the classical configuration without the inerter thus it has a gain margin about three times bigger than the classical control unit.

Fig. 5 shows the maximum stable gain of the feedback loop obtained from Eq. (14) as a function of the damping ratio of the inertial actuator ζ_2 and the inerter coefficient δ. The graph shows that the maximum stable gain increases as both the damping ratio and inerter coefficient increase. The use of an inerter in this application is particularly appealing in the light of the fact that a relatively large inertance can be achieved with a small increase of the total mass of the control unit [49].

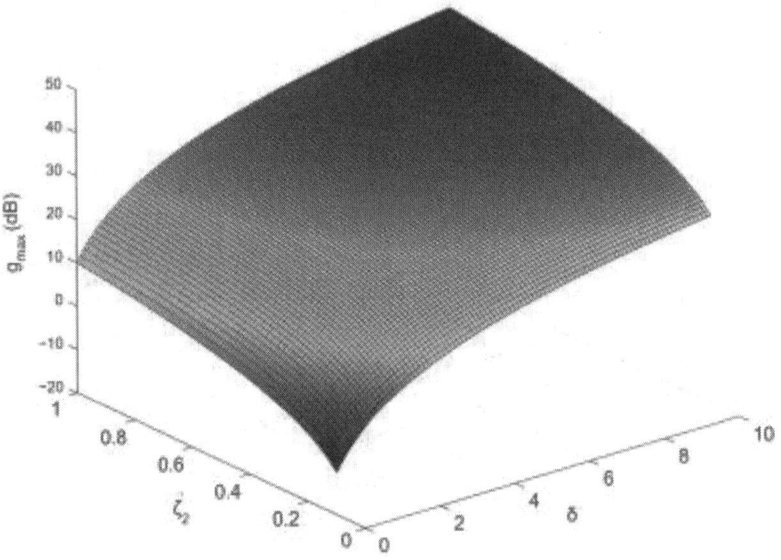

Figure. 5. Maximum stable gain as a function of the damping ratio ζ_2 and the inerter coefficient δ.

4.2. Performance

The performance of the controller is assessed in terms of the reduction of the kinetic energy of the system under control when excited by a broadband white noise force. The solid line in Fig. 6 shows the PSD of the kinetic energy of the primary system when the feedback gain, α, is zero.

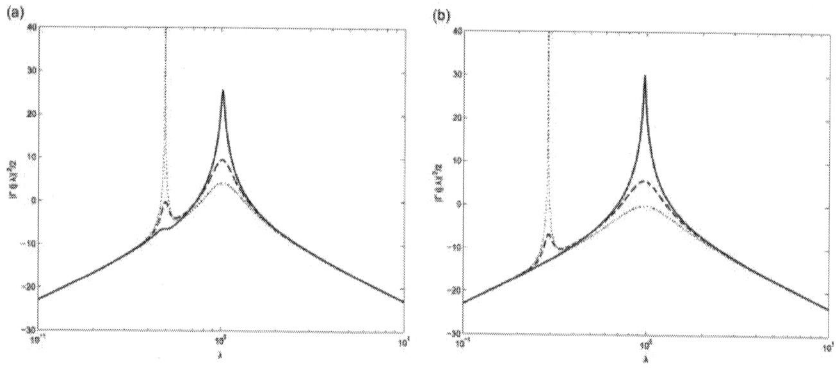

Figure. 6. PSD of the kinetic energy for zero gain (solid line) when the control gain that guarantees 6 dB gain margin (dashed line) and the maximum stable gain (dotted line) are implemented. Inertial actuator (a) without and (b) with the inerter δ=2.

Since the mass of the actuator is only 10% the mass of the primary system the spectrum of the kinetic energy is dominated by the resonant response of the primary system and the passive effect of the actuator is negligible.

As shown by the dashed line in Fig. 6, as the feedback gain is increased, significant attenuation is initially obtained at resonance frequency of the primary system for both configurations without (a) and with (b) the inerter.

At gains approaching the stability limit (dotted line), however, there is also a significant enhancement of the vibration at the natural frequency of the actuator, due to the positive feedback in this frequency region caused by the phase response of the actuator. Comparing Fig. 6(a) and (b) it can be noticed that for the control unit with the inerter the resonance of the actuator is shifted down in frequency thus, as shown in the previous subsection, the control system exhibits a greater gain margin which improves the control performance allowing a greater reduction in the spectrum of the kinetic energy of the primary system.

If the total kinetic energy of the primary system is calculated from Eq. (20) its variation with feedback gain, normalised by the kinetic energy with no control, is shown in Fig. 7 for the case with (solid line) and without (dashed line) the inerter. The feedback gain in Fig. 7has been normalised to the value of maximum stable gain. The total kinetic energy initially decreases as the feedback gain is increased, before increasing again as the maximum stable gain is approached. The minimum of the kinetic energy is reached when the feedback gain is set to the optimal value given by Eq. (21) and is shown by the circular markers. From the plot in Fig. 7 it can be clearly seen that when the inerter is added in parallel with the suspension of the inertial actuator additional 5 dB reduction in the kinetic energy of the primary system can be achieved compare to the case without the inerter when the gain is set to the optimal value.

To better understand the influence of the inerter on the performance of the control unit, the additional reduction in the kinetic energy of the primary system due to the effect of the inerter when the gain is set to the optimal value has been plotted against the inertance δin Fig. 8.

The plot shows that a rapid increase in the performance of the control system can be achieved for relatively low values of the inertance δ while for values greater than 20 only a moderate improvement in the performance can be obtained. The circular marker in Fig. 8 shows the value of δ considered the simulations.

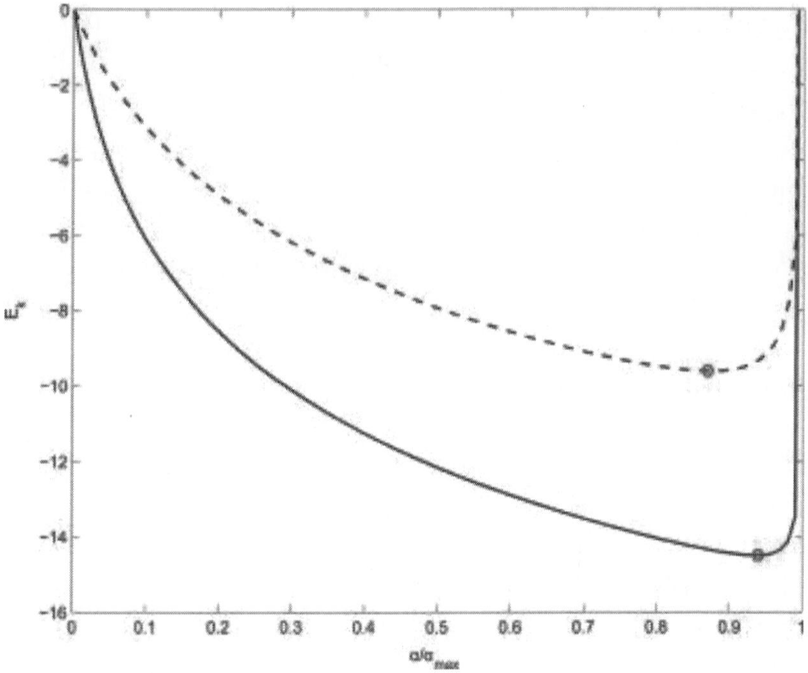

Figure. 7. Reduction in dB of the kinetic energy of the primary system as a function of the control gain g normalised to the maximum stable gain for the control unit with the inerter $b_2 = 0.2$ **kg** (solid line) and without (dashed line) inerter.

As already pointed out in the previous section an alternative control strategy to tune the gain α is to maximise the power absorbed by the controller. Fig. 9 shows the power absorbed by the control unit, I_p, as a function of the normalised feedback gain for the configuration with (solid line) and without (dashed line) the inerter. Fig. 9 shows that when the control gain is zero some power is absorbed by the passive damper of the suspension. As the control gain is increased more and more power is absorbed until a maximum is reached. Comparing the solid and dashed line in Fig. 9 it can be seen that the control unit with the inerter is able to absorb more power than the classical configuration without the inerter. As also analytically demonstrated in the previous section, comparing Fig. 7 and Fig. 9 it can be notice that maximising the absorbed power results in the same control gain as minimising the kinetic energy. The control unit could thus be made self-tuneable if an algorithm that adjusts the control gain to maximise the absorbed power is employed [43].

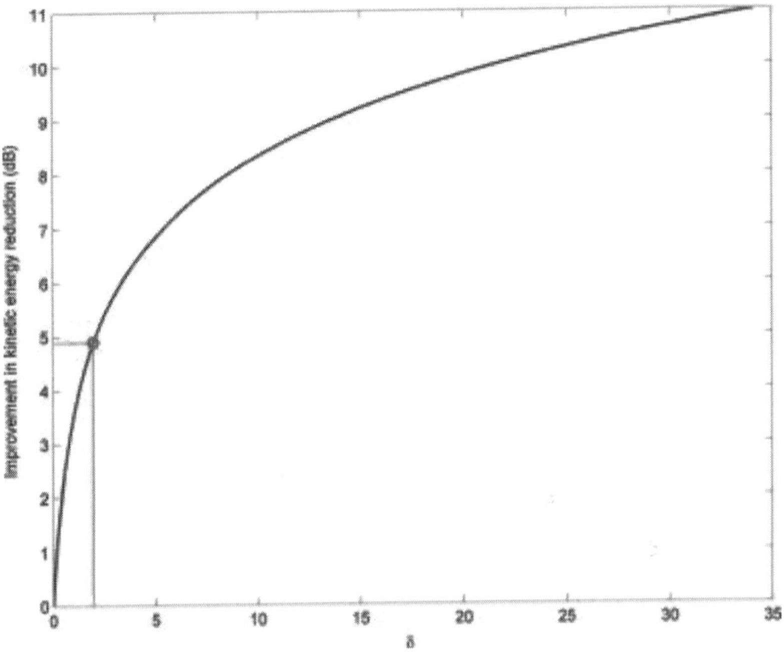

Figure. 8. Improvement in the kinetic energy reduction as a function of δ. The circular marker shows the value of δconsidered the simulations.

When the control gain approaches the maximum stable value the active force injects power into the system that is then dissipated by the passive damper c_2. This phenomenon becomes more evident when the ratio between m_2 and m_1 decreases. Further investigation on the power flows in the system is currently under investigation.

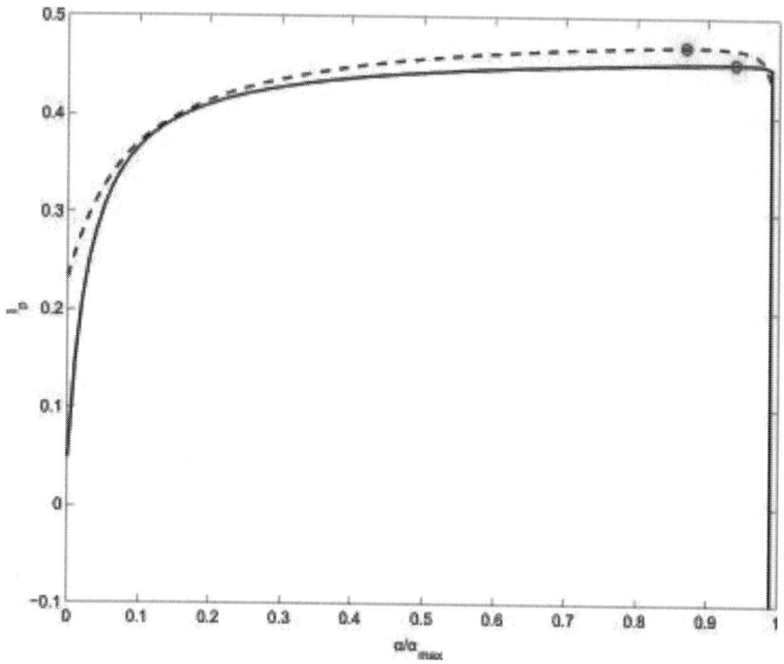

Figure. 9. The power absorbed by the control unit as a function of the control gain g normalised to the maximum stable gain for the control unit with the inerter $b_2 = 0.1$ kg (solid line) and without (dashed line) inerter. The two circular markers show the locations of the maxima.

5. CONCLUSIONS

This paper presents a theoretical study on the implementation of a velocity feedback loop to reduce the vibration of a structure modelled as a single degree of freedom system. The control unit consist of an inertial actuator with a mass suspended on a spring, a damper and an inerter with an electromechanical transducer in parallel with the suspension. The actuator is fed with a signal proportional to the velocity measured by a sensor located at the footprint of the control unit. It is shown that the use of the inerter has the effect of shifting down the fundamental resonance frequency of the control unit. This allows the use of a relatively stiff spring suspension able to support the static load of the inertial mass and at the same time improve stability and performance of the control system. The study also consider the tuning of the feedback control gain in order to either minimise the kinetic energy of the system under control or maximise the power absorbed by the controller. It has been analytically demonstrated that the two tuning strategies are equivalent.

ACKNOWLEDGEMENTS

This work was supported by the EPSRC through "Engineering Nonlinearity Project" programme grant (EP/K003836/1).

REFERENCES

1. H. Frahm, Device for damping vibrations of bodies, U.S. Patent, 989,958, 1911.

2. J.P. Den Hartog, Mechanical Vibrations, 4th ed. McGraw-Hill, New York, 1956.

3. T. Asami, O. Nishihara, A.M. Baz, Analytical solutions to H1 and H2 optimization of dynamic vibration absorbers attached to damped linear systems, Journal of Vibration and Acoustics 124 (2002) 284–295.

4. S.H. Crandall, W.D. Mark, Random Vibration in Mechanical Systems, Academic Press, New York, 1963.

5. Y. Iwata, On the construction of the dynamic vibration absorber, Japanese Society of Mechanical Engineering 820 (1982) 150–152.

6. S. Krenk, Frequency analysis of the tuned mass damper, Journal of Applied Mechanics 72 (2005) 936–942.

7. O. Nishihara, T. Asami, Closed-form solutions to the exact optimizations of dynamic vibration absorbers (minimizations of the maximum amplitude magnification factors), Journal of Vibration and Acoustics 124 (2002) 576–582.

8. O. Nishihara, H. Matsuhisa, Design of a dynamic vibration absorber for minimisation of maximum amplitude magnification factor (derivation of algebric exact solution), Transactions of the Japan Society of Mechanical Engineers, Series C 62–614 (1997) 3438–3445.

9. G.B. Warburton, Optimum absorber parameters for various combinations of response and excitation parameters, Journal of Earthquake Engineering and Structural Dynamics 10 (1982) 381–401.

10. H. Yamaguchi, Damping of transient vibration by a dynamic absorber, Transactions of the Japan Society of Mechanical Engineers, Series C 54 (1988) 561–568.

11. M. Zilletti, S.J. Elliott, E. Rustighi, Optimisation of dynamic vibration absorbers to minimise kinetic energy and maximise internal power dissipation, Journal of Sound and Vibration 331 (2012) 4093–4100.

12. L. Zuo, S.A. Nayfeh, Minimax optimization of multi-degree-of-freedom tuned-mass dampers, Journal of Sound and Vibration 272 (2004) 893–908.

13. P. Gardonio, M. Zilletti, Integrated tuned vibration absorbers: a theoretical study, The Journal of the Acoustical Society of America 134 (2013) 3631–3644.

14. P. Gardonio, M. Zilletti, Sweeping tuneable vibration absorbers for low-mid frequencies vibration control, Journal of Sound and Vibration 354 (2015) 1–12

15. M. Zilletti, P. Gardonio, Experimental implementation of switching and sweeping tuneable vibration absorbers for broadband vibration control, Journal of Sound and Vibration 334 (2015) 164–177.

16. F. Fahy, P. Gardonio, Sound and Structural Vibration, Radiation, Transmission and Response, 2nd ed. Academic Press, Oxford, UK, 2007.

17. S.J. Elliott, Global vibration control through local feedback, in: D. Wagg, I. Bond, P. Weaver, M. Friswell (Eds.), Adaptive Structure: Engineering Applications, John Wiley & Sons Ltd., Chichester2007, pp. 59–87.

18. S.J. Elliott, P. Gardonio, T.C. Sors, M.J. Brennan, Active vibroacoustic control with multiple local feedback loops, The Journal of the Acoustical Society of America 111 (2002) 908–915.

19. C. González Díaz, P. Gardonio, Feedback control laws for proof-mass electrodynamic actuators, Smart Materials and Structures 16 (2007) 1766.

20. Preumont, Vibration Control of Active Structures, Kluwer Academic, London, 2002.

21. S.J. Elliott, M. Serrand, P. Gardonio, Feedback stability limits for active isolation systems with reactive and inertial actuators, Transactions of the ASME, Journal of Vibration and Acoustics 123 (2001) 250–261.

22. L. Benassi, S.J. Elliott, Active vibration isolation using an inertial actuator with local displacement feedback control, Journal of Sound and Vibration 278 (2004) 705–724.

23. C. Paulitsch, P. Gardonio, S.J. Elliott, Active vibration control using an inertial actuator with internal damping, The Journal of the Acoustical Society of America 119 (2006) 2131–2140.

24. J. Rohlfing, S.J. Elliott, P. Gardonio, Feedback compensator for control units with proof-mass electrodynamic actuators, Journal of Sound and Vibration 331 (2012) 3437–3450.

25. M.C. Smith, Synthesis of mechanical networks: the inerter, IEEE Transactions on Automatic Control 47 (2002) 1648–1662.

26. Y. Hu, M.Z.Q. Chen, Performance evaluation for inerter-based dynamic vibration absorbers, International Journal of Mechanical Sciences 99 (2015) 297–307.

27. P. Brzeski, E. Pavlovskaia, T. Kapitaniak, P. Perlikowski, The application of inerter in tuned mass absorber, International Journal of Non-Linear Mechanics 70 (2015) 20–29.

28. P. Brzeski, T. Kapitaniak, P. Perlikowski, Novel type of tuned mass damper with inerter which enables changes of inertance, Journal of Sound and Vibration 349 (2015) 56–66.

29. M.C. Smith, F.-C. Wang, Performance benefits in passive vehicle suspensions employing inerters, Vehicle System Dynamics 42 (2004) 235–257.

30. M.Z.Q. Chen, C. Papageorgiou, F. Scheibe, W. Fu-cheng, M.C. Smith, The missing mechanical circuit element, IEEE Circuits and Systems Magazine 9 (2009) 10–26.

31. Y. Shen, L. Chen, X. Yang, D. Shi, J. Yang, Improved design of dynamic vibration absorber by using the inerter and its application in vehicle suspension, Journal of Sound and Vibration 361 (2016) 148–158.

32. P. Li, J. Lam, K.C. Cheung, Control of vehicle suspension using an adaptive inerter, Proceedings of the Institution of Mechanical Engineers, Part D: Journal of Automobile Engineering (2015).

33. M.Z.Q. Chen, K. Wang, S. Zhan, L. Chanying, Realizations of a special class of admittances with strictly lower complexity than

canonical forms, IEEE Transactions on Circuits and Systems I: Regular Papers 60 (2013) 2465–2473.

34. M.Z.Q. Chen, K. Wang, Y. Zou, J. Lam, Realization of a special class of admittances with one damper and one inerter for mechanical control, IEEE Transactions on Automatic Control 58 (2013) 1841–1846.

35. K. Wang, M.Z.Q. Chen, Y. Hu, Synthesis of biquadratic impedances with at most four passive elements, Journal of the Franklin Institute 351 (2014) 1251–1267.

36. S. Evangelou, D.J.N. Limebeer, R.S. Sharp, M.C. Smith, Control of motorcycle steering instabilities, IEEE Control Systems 26 (2006) 78–88.

37. S. Evangelou, D.J.N. Limebeer, R.S. Sharp, M.C. Smith, Mechanical steering compensators for high-performance motorcycles, Journal of Applied Mechanics 74 (2006) 332–346.

38. F.-C. Wang, M.-K. Liao, B.-H. Liao, W.-J. Su, H.-A. Chan, The performance improvements of train suspension systems with mechanical networks employing inerters, Vehicle System Dynamics 47 (2009) 805–830.

39. J.Z. Jiang, A.Z. Matamoros-Sanchez, R.M. Goodall, M.C. Smith, Passive suspensions incorporating inerters for railway vehicles, Vehicle System Dynamics 50 (2012) 263–276.

40. F.-C. Wang, M.-F. Hong, C.-W. Chen, Building suspensions with inerters, Proceedings of the Institution of Mechanical Engineers, Part C: Journal of Mechanical Engineering Science 224 (2010) 1605–1616.

41. X. Dong, Y. Liu, M.Z.Q. Chen, Application of inerter to aircraft landing gear suspension, Proceedings of the 34th Chinese Control Conference (CCC), 2015, pp. 2066–2071.

42. M.Z.Q. Chen, Y. Hu, L. Huang, G. Chen, Influence of inerter on natural frequencies of vibration systems, Journal of Sound and Vibration 333 (2014) 1874–1887.

43. M. Zilletti, S.J. Elliott, P. Gardonio, Self-tuning control systems of decentralised velocity feedback, Journal of Sound and Vibration 329 (2010) 2738–2750.

44. M. Zilletti, S.J. Elliott, P. Gardonio, E. Rustighi, Experimental implementation of a self-tuning control system for decentralised velocity feedback, Journal of Sound and Vibration 331 (2012) 1–14.

45. M. Zilletti, P. Gardonio, S.J. Elliott, Optimisation of a velocity feedback controller to minimise kinetic energy and maximise power dissipation, Journal of Sound and Vibration 333 (2014) 4405–4414.

46. D.E. Newland, An Introduction to Random Vibrations, Spectral & Wavelet Analysis, Third edition ed. Dover Publications, Inc., Mineola, New York, 1975.

47. S.J. Elliott, M. Zilletti, P. Gardonio, Apparatus and method of vibration control, Patent, WO2011114165 A1, 2011.

48. R.S. Langley, A general mass law for broadband energy harvesting, Journal of Sound and Vibration 333 (2014) 927–936.

49. Y. Hu, M.Z.Q. Chen, Z. Shu, L. Huang, Analysis and optimisation for inerter-based isolators via fixed-point theory and algebraic solution, Journal of Sound and Vibration 346 (2015) 17–36.

Index

O

P

S

T

V